Despeckle Filtering for Ultrasound Imaging and Video

Volume I

Algorithms and Software

Second Edition

Synthesis Lectures on Algorithms and Software in Engineering

Editor
Andreas Spanias, *Arizona State University*

Despeckle Filtering for Ultrasound Imaging and Video, Volume I: Algorithms and Software, Second Edition
Christos P. Loizou and Constantinos S. Pattichis
2015

Latency and Distortion of Electromagnetic Trackers for Augmented Reality Systems
Henry Himberg and Yuichi Motai
2014

Bandwidth Extension of Speech Using Perceptual Criteria
Visar Berisha, Steven Sandoval, and Julie Liss
2013

Control Grid Motion Estimation for Efficient Application of Optical Flow
Christine M. Zwart and David H. Frakes
2013

Sparse Representations for Radar with MATLAB ™ Examples
Peter Knee
2012

Analysis of the MPEG-1 Layer III (MP3) Algorithm Using MATLAB
Jayaraman J. Thiagarajan and Andreas Spanias
2011

Theory and Applications of Gaussian Quadrature Methods
Narayan Kovvali
2011

Algorithms and Software for Predictive and Perceptual Modeling of Speech
Venkatraman Atti
2011

Adaptive High-Resolution Sensor Waveform Design for Tracking
Ioannis Kyriakides, Darryl Morrell, and Antonia Papandreou-Suppappola
2010

MATLAB™ Software for the Code Excited Linear Prediction Algorithm: The Federal Standard-1016
Karthikeyan N. Ramamurthy and Andreas S. Spanias
2010

OFDM Systems for Wireless Communications
Adarsh B. Narasimhamurthy, Mahesh K. Banavar, and Cihan Tepedelenliouglu
2010

Advances in Modern Blind Signal Separation Algorithms: Theory and Applications
Kostas Kokkinakis and Philipos C. Loizou
2010

Advances in Waveform-Agile Sensing for Tracking
Sandeep Prasad Sira, Antonia Papandreou-Suppappola, and Darryl Morrell
2008

Despeckle Filtering Algorithms and Software for Ultrasound Imaging
Christos P. Loizou and Constantinos S. Pattichis
2008

Despeckle Filtering for Ultrasound Imaging and Video, Volume I: Algorithms and Software, Second Edition
Christos P. Loizou and Constantinos S. Pattichis

ISBN: 978-3-031-00395-0   paperback
ISBN: 978-3-031-01523-6   ebook

DOI 10.1007/978-3-031-01523-6

A Publication in the Springer series
SYNTHESIS LECTURES ON ALGORITHMS AND SOFTWARE IN ENGINEERING

Lecture #14
Series Editor: Andreas Spanias, *Arizona State University*
Series ISSN
Print 1938-1727   Electronic 1938-1735

Despeckle Filtering for Ultrasound Imaging and Video
Volume I
Algorithms and Software

Second Edition

Christos P. Loizou
School of Sciences and Engineering, Intercollege, Cyprus

Constantinos S. Pattichis
University of Cyprus

SYNTHESIS LECTURES ON ALGORITHMS AND SOFTWARE IN ENGINEERING #14

ABSTRACT

It is well known that speckle is a multiplicative noise that degrades image and video quality and the visual expert's evaluation in ultrasound imaging and video. This necessitates the need for robust despeckling image and video techniques for both routine clinical practice and tele-consultation. The goal for this book (book 1 of 2 books) is to introduce the problem of speckle occurring in ultrasound image and video as well as the theoretical background (equations), the algorithmic steps, and the MATLAB™ code for the following group of despeckle filters: linear filtering, nonlinear filtering, anisotropic diffusion filtering, and wavelet filtering. This book proposes a comparative evaluation framework of these despeckle filters based on texture analysis, image quality evaluation metrics, and visual evaluation by medical experts. Despeckle noise reduction through the application of these filters will improve the visual observation quality or it may be used as a pre-processing step for further automated analysis, such as image and video segmentation, and texture characterization in ultrasound cardiovascular imaging, as well as in bandwidth reduction in ultrasound video transmission for telemedicine applications. The aforementioned topics will be covered in detail in the companion book to this one. Furthermore, in order to facilitate further applications we have developed in MATLAB™ two different toolboxes that integrate image (IDF) and video (VDF) despeckle filtering, texture analysis, and image and video quality evaluation metrics. The code for these toolsets is open source and these are available to download complementary to the two books.

KEYWORDS

speckle, despeckle, noise filtering, ultrasound, ultrasound imaging, ultrasound video, cardiovascular imaging and video, SAR, texture, image quality, video quality, carotid artery

To my Family

Christos P. Loizou

To my Family

Constantinos S. Pattichis

"Show thyself in all things an example of good works,
in teaching, in integrity and dignity;
let thy speech be sound and blameless,
so that anyone opposing may be put to shame,
having nothing bad to say for us.
Exhort slaves to obey their masters,
pleasing them in all things and not opposing them."

Titus 2:7–9.

Contents

Preface

Speckle can be modeled as a multiplicative noise source that degrades image and video quality and the visual evaluation in ultrasound and SAR imaging. This necessitates the need for robust despeckling techniques in a wide spectrum of the aforementioned imaging applications. Despeckle filtering is a rapidly emerging research area with several applications for ultrasound images and videos. The goal for this book is to introduce the theoretical background (equations), the algorithmic steps, and the MATLAB™ code for the following group of despeckle filters: linear filtering, nonlinear filtering, anisotropic diffusion filtering, and wavelet filtering. The filters covered represent only a snapshot of the vast number of despeckle filters published in the literature. Moreover, selected representative applications of image despeckling covering a variety of ultrasound image processing tasks are presented. Most importantly, a despeckle filtering and evaluation protocol is proposed based on texture analysis, image quality evaluation metrics, and visual evaluation by experts. The source code of the algorithms presented in this book has been made available on the web, thus enabling researchers to more easily exploit the application of despeckle filtering in their problems under investigation.

This book is organized into eight chapters. Chapter 1 presents a brief overview of ultrasound imaging and video, speckle noise, modeling, and filtering. Chapter 2 covers the evaluation methodology based on texture and statistical analysis, image quality evaluation metrics, and the experiments carried out for visual evaluation. The theoretical background (equations), the algorithmic steps, and the MATLAB™ code of selected despeckle filters are presented for linear despeckle filtering, for nonlinear despeckle filtering, for diffusion despeckle filtering and for wavelet despeckle filtering in Chapters 3–6, respectively. Chapter 4 presents the applications of despeckle filtering techniques in ultrasound images of the carotid and cardiac ultrasound images. Chapter 7 discusses, compares, and evaluates the proposed despeckle filtering techniques. Chapter 8 presents the summary and future directions, where a despeckling filtering protocol is also proposed. Finally, at the end of this book, an Appendix provides details about two different despeckle filtering MATLAB™ toolboxes for ultrasound imaging and video of the carotid artery.

This book is intended for all those working in the field of image and video processing technologies, and more specifically in medical imaging and in ultrasound image and video preprocessing and analysis. It provides different levels of material to researchers, biomedical engineers, computing engineers, and medical imaging engineers interested in developing imaging systems with better quality images, limiting the corruption of speckle noise.

We hope that this book will be a useful reference for all the readers in this important field of research and that it will contribute to the development and implementation of innovative imaging and video systems, enabling the provision of better quality images.

Christos P. Loizou and Constantinos S. Pattichis
April 2015

Acknowledgments

We wish to thank all the members of our ultrasound imaging team for the long discussions, advice, encouragement, and constructive criticism they provided to us during the course of this research work. First of all, we would like to express our sincere thanks to Emeritus Prof. Andrew Nicolaides, of the Faculty of Medicine, Imperial College of Science, Technology and Medicine, UK, and founder of the Vascular Screening and Diagnostic Centre in Cyprus. Furthermore, we would like to express our thanks to Dr. Marios Pantziaris, consultant neurologist, at the Cyprus Institute of Neurology and Genetics; Dr. Theodosis Tyllis, consultant physician in the private sector in Cyprus; Associate Professor Efthyvoulos Kyriakou, at the Frederick University, Cyprus; Dr. Christodoulos Christodoulou, Research Associate at the University of Cyprus; and Professor Marios Pattichis, University of New Mexico, USA. Last, but not least, we would like to thank Prof. Andreas Spanias, Arizona State University, USA, for his proposal and encouragement in writing this book, and Joel Claypool, and the rest of the staff at Morgan & Claypool, for their understanding, patience, and support in materializing this project.

Christos P. Loizou and Constantinos S. Pattichis
April 2015

List of Symbols

$a_{i,j}$	Additive noise component on pixel i, j
$\alpha_{comp}, \beta_{comp}$	Logarithmic compression parameters
$\beta(s)$	Snake stiffness of the energy functional
β_{GVF}	GVF snake rigidity parameter
C	Speckle Index
$CV\%$	Coefficient of variation
$cd(\|\nabla g\|), c_{i,j}$	Diffusion coefficient
c_{adsr}	Speckle reducing anisotropic diffusion coefficient
c	Constant controlling the magnitude of the potential
$c_{s\sin\_1}, c_{s\sin\_2}$	Constants used to calculate the SSIN
c_2	Positive weighting factor
Γ	Number of directions, which diffusion is computed
γ	Signal-to-noise ratio (SNR)
$D \in \Re^{2x2}$	Symmetric positive semi-definite diffusion tensor representing the required diffusion in both gradient and contour directions
D_f	Fractal dimension
D	Matrix used to calculated the image energy of the snake, $E_{image}(v)$
$D_{viewing}$	Viewing distance
DR	Dynamic range of input ultrasound signal
$d(k)$	Wavelet coefficient for the wavelet filtering
Δf	Frequency shift (Doppler frequency shift)
Δr	Distance between two pixels
∇g	The gradient magnitude of image $g(x, y)$(gradient)
$\nabla g_{i,j}$	Directional derivative (simple difference) at location i, j
$f_1 \ldots f_{13}$	SGLDM texture measures from Haralick
$f_x(x, y)$	First order differential of the edge magnitude along the x-axis
$f_{i,j}$	Noise-free signal ultrasound signal in discrete form (the new image) on pixel i, j
v	Frequency of ultrasound wave
f_0	Transmitted frequency of ultrasound signal
$feat\_dis_i$	Percentage distance
$g_{i,j}$	Observed ultrasound signal in discrete formulation after logarithmic compression

$g(x, y)$	Observed ultrasound signal after logarithmic compression, representing image intensity at location (x, y)
G	Linear gain of the amplifier
$G\sigma * g_{i,j}$	Image convolved with Gaussian smoothing filter
$G\sigma$	Gaussian smoothing filter
$\bar{g}_i, \bar{f}_i$	Mean gravity of the searching pixel region in image g or f
g_{max}, g_{min}	Maximum and minimum gray level values in a pixel neighborhood
Hz, KHz, MHz	Hertz, Kilohertz, Megahertz
HX, HY	Entropies of p_x and p_y
$H^{(k)}$	Hurst coefficients
$H(x, y)$	Array of points of the same size for the HT
HD	Hausdorff distance
η_s	Spatial neighborhood of pixel i, j
$|\eta_s|$	Number of neighbors (usually four except at the image boundaries)
θ_i	Phase shift relative to the insonated ultrasound wave
θ	Angle between the direction of movement of the moving object and the ultrasound beam
I	Identity matrix
$I_0(x)$	Modified Bessel function of the first kind of order 0
IMT_{mean}	Mean value of the IMT
IMT_{min}	IMT minimum value
IMT_{max}	IMT maximum value
IMT_{median}	IMT median value
k	Coefficient of variation for speckle filtering
λ	Wavelength of ultrasound wave
λ_π	Lai & Chin snake energy regularization parameter, $E_{snake}(v)$
$\lambda_d \in \Re^+$	Rate of diffusion for the anisotropic diffusion filter
m_{i1}, m_{i2}	Mean values of two classes (asymptomatic, symptomatic)
$cm/s, cm/s$	Meters per second, centimeters per second
μ	Mean
N	Number of scatterers within a resolution cell
N_{feat}	Number of features in the feature set
$n_{i,j}$	Multiplicative noise component (independent of $g_{i,j}$, with mean 0) on pixel i, j
$nl_{i,j}$	Multiplicative noise component after logarithmic compression on pixel i, j
$n(s)$	Normal force tensor
ξ_i	Amount of ultrasound signal backscattered by scatterer $W2LOK$

$p_x(i)$	*ith* entry in the marginal probability matrix obtained by summing the rows of $p(i, j)$
Q	Mathematically defined universal quality index
$R = 1 - \frac{1}{1+\sigma^2}$	Smoothness of an image
$Score\_Dis$	Score distance between two classes (asymptomatic, symptomatic)
$s_e = \sigma_{IMT}/\sqrt{2}$	Inter-observer error
s_{max}	Maximum pixel value in the image
s^2	Structural energy
σ_{IMT}	IMT standard deviation
σ_{fg}	Covariance between two images f and g
σ	Standard deviation
σ^2	Variance
σ^3	Skewness
σ^4	Kurtosis
$2\sigma^2$	Diffuse energy
σ_n	Standard deviation of the noise
σ_w^2	Variance of the gray values in a pixel window

List of Abbreviations

ACSRS	Asymptomatic Carotid Stenosis
DsFad	Perona and Malik anisotropic diffusion filter
DsFadsr	Speckle-reducing anisotropic diffusion filter
ASM	Angular second moment
ATL HDI-3000	ATL 3000 ultrasound scanner
ATL HDI-5000	ATL 5000 ultrasound scanner
DsFca	Linear scaling of the gray-levels despeckle filter
CAT	Computer-assisted tomography
CCA	Common carotid artery
CSR	Contrast-to-speckle ratio
CT	Computer tomography
CW	Continuous wave
DR	Dynamic range
DS	Despeckled
DSCQS	Double stimulus continuous quality scale
DSIS	Double stimulus impairment scale
DVD	Digital video
DWT	Discrete wavelet transform
E	Effectiveness measure
ECA	External carotid artery
ECST	European carotid surgery trial
Err	Error summation in the form of the Minkowski metric
FDTA	Fractal dimension texture analysis
FFT	Fast Fourier transform
FPS	Fourier power spectrum
GAE	Geometric average error
GF	Geometric filtering
DsFgf4d	Geometric despeckle filter
DsFgfminmax	Geometric despeckle filter utilizing minimum maximum values
GGVF	Generalized gradient vector flow
GLDS	Gray level difference statistics
GVF	Gradient vector flow
HD	Hausdorff distance

HF	Maximum homogeneity
HM	Homomorphic
DsFhomo	Homomorphic despeckle filter
DsFhomog	Most homogeneous neighborhood despeckle filter
HVS	Human visual system
ICA	Internal carotid artery
IDM	Inverse difference moment
IDV	Intensity difference vector
IMC	Intima-media complex
IMT	Intima-media thickness
IVUS	Intravascular ultrasound
kNN	The statistical k-nearest-neighbor classifier
DsFlecasort	Linear scaling and sorting despeckle filter
LS	Linear scaling
DsFls	Linear scaling of the gray level values despeckle filter
DsFlsmedcd	Lee diffusion despeckle filter
DsFlsminsc	Minimum speckle index homogeneous mask despeckle filter
DsFlsminv1d	Minimum variance homogeneous 1D mask despeckle filter
DsFlsmv	Mean and variance local statistics despeckle filter
DsFlsmvsk2d	Mean variance, higher-moments local statistics despeckle filter
DsFlsmvske1d	Mean, variance, skewness, kurtosis 1D local statistics despeckle filter
M	Manual
DsFmedian	Median despeckle filter
DsFhmedian	Hybrid median filter
MF	Multi-resolution fractal
MMSE	Minimum mean-square error
MN	Manual normalized
MRI	Magnetic resonance imaging
MSE	Mean square error
N	Normalized
ND	Normalized despeckled
NE	North east
NF	No filtering
NGTDM	Neighbourhood gray tone difference matrix
NIE	Normalized image energy
DsFnldif	Nonlinear coherent diffusion despeckle filter
NS	Not significant difference
NST	North south
NTSE	Normalized total snake energy

P	Precision
PDE	Partial differential equation
PDF	Probability density function
PET	Positron emission tomography
PSNR	Peak signal-to-noise ratio
PW	Pulsed wave
R	Sensitivity (or recall)
RF	Radio frequency
RMSE	Root mean square error
ROC	Receiver operating characteristic
S	Significant difference
Sp	Specificity
SAR	Synthetic aperture radar
SD	Standard deviation
SE	South east
SFM	Statistical feature matrix
SGLDM	Spatial gray level dependence matrices
SGLDMm	Spatial gray level dependence matrix mean values
SGLDMr	Spatial gray level dependence matrix range of values
SNR	Signal-to-noise ratio
SPECT	Single photon emission computer tomography
SSIN	Structural similarity index
TEM	Laws texture energy measures
TGC	Time gain compensation
TIA	Transient ischemic attacks
TV	Television
$DsFwaveltc$	Wavelet despeckle filter
WE	West east
$DsFwiener$	Wiener despeckle filter
WN	West north
WS	West south
WT	Wavelet transform
β_{err}	Minkowski error coefficient
1D	One-dimensional
2D	Two-dimensional
3D	Three-dimensional

CHAPTER 1

Introduction to Speckle Noise in Ultrasound Imaging and Video

According to an old Chinese proverb, "a picture is worth a thousand words." In the modern age, this concept is still true for computer vision and image processing tasks, where we aim to develop and implement better systems and tools that give us different perspectives on the same image thus allowing us to understand not only its content, but also its meaning and significance. Image processing cannot compete with the human eye in terms of accuracy but it can perform better on observational consistency and ability to carry out detailed mathematical operations. In the course of time, image-processing research has evolved from basic low-level pixel operations to high-level analysis that now includes sophisticated techniques for image interpretation and analysis. These new techniques are being developed in order to gain a better understanding of images based on the relationships between its components, context, history, and knowledge gained from a range of sources.

The purpose of this chapter is to give a brief overview of ultrasound imaging and video, and present its basic principles and limitations. Furthermore, speckle noise is introduced as a major noise factor, which limits image and video resolution and hinders further processing analysis in ultrasound images and videos. We then introduce different despeckle filtering techniques that may be applied as a pre-processing step for denoising of ultrasound images and videos. Moreover, at the end of the chapter, we present the physical properties of speckle noise, its mathematical model, and its limitations. Finally, a few examples of despeckle filtering for real ultrasound images and videos are given and some of its limitations are discussed.

1.1 A BRIEF REVIEW OF ULTRASOUND IMAGING AND VIDEO

Medical imaging technology has experienced a dramatic change in the last 30 years. Previously, only X-ray radiographs were available, which showed the organs as shadows on photographic film. With the advent of modern computers and digital imaging technology, new imaging modalities like computer tomography (CT or CAT computer-assisted tomography), magnetic resonance imaging (MRI), positron emission tomography (PET), and ultrasound, which deliver cross-sectional images and videos of a patient's anatomy and physiology, have been developed. Among the imaging techniques employed are X-ray angiography, X-ray, CT, ultrasound imaging, MRI, PET, and single photon emission computer tomography (SPECT). MRI and CT have advantages compared to ultrasound, in the sense that higher resolution and clearer images and videos are produced.

Imaging and video techniques have long been used for assessing and treating cardiac [1] and carotid disease [2–4]. Today's available imaging modalities produce a wide range of image data types for disease assessment which includes, 2D projection images and videos, reconstructed three-dimensional (3D) images and videos, 2D slice images, true 3D images, time sequences of 2D and 3D images, and sequences of 2D interior view (endoluminal) images. Videos can be generated in the same manner as images, by producing a series of images acquired continuously at a specific frame rate.

The use of ultrasound in the diagnosis and assessment of imaging organs and soft tissue structures as well as human blood, is well established [5] (see Fig. 1.1, illustrating imaging scanners). Because of its non-invasive nature and continuing improvements in imaging quality, ultrasound imaging is progressively achieving an important role in the assessment and characterization of cardiac imaging (see Fig. 1.2), and the assessment of carotid artery disease [4], [6]–[13] (see Fig. 1.3). The main disadvantage of ultrasound is that it does not work well in the presence of bone or gas, and the operator needs a high level of skill in both image acquisition and interpretation to carry out the clinical evaluation. On the other hand, standard angiography cannot give reliable information on the cross-sectional structure of the arteries [7]. This makes it difficult to accurately assess the build-up of plaque along the artery walls. B-mode ultrasound imaging or intravascular ultrasound (IVUS) has emerged and it is widely used for visualizing carotid plaques and the assessment of plaque characteristics related to the onset of neurological symptoms. IVUS needs the insertion of a catheter into a vessel of interest that is equipped with an ultrasonic transducer enabling the reproduction of real-time cross sectional images. However, reproducible measurements of the severity of the plaque in 2D and 3D ultrasound are made difficult because of the complex shapes, asymmetry of carotid plaques, and the speckle noise present in ultrasound images [5] as well as in videos [14, 15]. Furthermore, IVUS is an invasive method, as a catheter is inserted in the artery under investigation and possesses therefore, a certain risk for the patient.

The use of ultrasound in medicine began during the second world war in various centres around the world. The work of Dr. Karl Theodore Dussik in Austria in 1942 [16] on ultrasound transmission investigating the brain provides the first published work on medical ultrasonics. Furthermore, although other researchers in the U.S., Japan, and Europe have also been cited as pioneers, the work of Professor Ian Donald [17] and his colleagues in Glasgow, in the mid-1950s, did much to facilitate the development of practical ultrasound technology and applications. This lead to the wider use of ultrasound in medical practice in subsequent decades.

From the mid-1960s onward, the advent of commercially available systems allowed the wider dissemination of the use of ultrasound. Rapid technological advances in electronics and

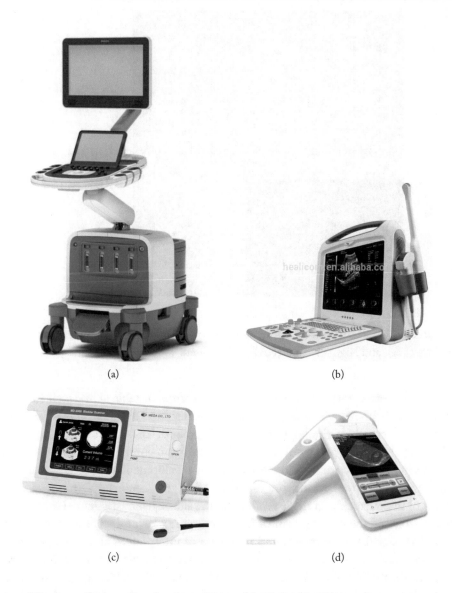

(a)

(b)

(c)

(d)

Figure 1.1: Ultrasound imaging and video systems: (a) Philips™ EPIQ 7 (http://www.healthca re.philips.com/main/products/ultrasound/systems/epiq7), (b) HY2000 4D portable color doppler ultrasound diagnostic system (Nanjing Healicom Medical Equipment Co., Ltd.), (c) Meda™ Co. Ltd. MD-6000 Portable bladder 3D/4D video ultrasound scanner (http://www.medicalexpo. com), and (d) Mobisante™ Inc., MobiUS portable ultrasound scanner that plugs into smartphones (http://www.mobisante.com).

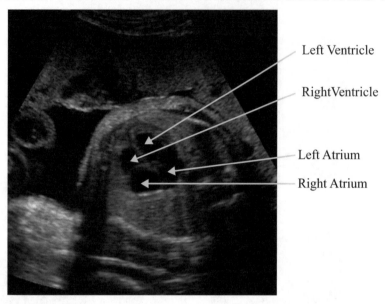

Left Ventricle

RightVentricle

Left Atrium

Right Atrium

Figure 1.2: Ultrasound B-mode cardiac image, where the Left Ventricle (LV), Right Ventricle (RV), Left Atrium (LA), and Right Atrium (RA) are indicated.

piezoelectric materials provided further improvements from bistable to grayscale images and from still images to real-time moving images (videos). The technical advances at this time (mid-1960s) led to the rapid growth in the applications of ultrasound. The development of Doppler ultrasound [18] has been progressing alongside the imaging technology but the fusing of the two technologies in Duplex scanning and the subsequent development of color Doppler imaging provided even more scope for investigating the circulation and blood supply to organs, tumors, etc. The advent of the microchip in the seventies and the subsequent exponential increase in processing power facilitated the development of faster and more powerful systems incorporating digital beam forming, signal enhancement, and new ways of interpreting and displaying data, such as power Doppler [18] and 3D imaging [19]. Ultrasound has long been recognized as a powerful tool for use in the diagnosis and evaluation of many clinical entities. Over the past decade, as higher quality and less expensive scanners are widely available, ultrasound has proliferated throughout various specialties.

Table 1.1 tabulates the human organs where ultrasound imaging can be applied as well as the different clinical conditions that can be diagnosed using ultrasound and other imaging modalities. Furthermore, Table 1.2 gives an overview of the different modalities that are degraded by speckle noise (see Section 1.2 Speckle Noise) and other types of noise.

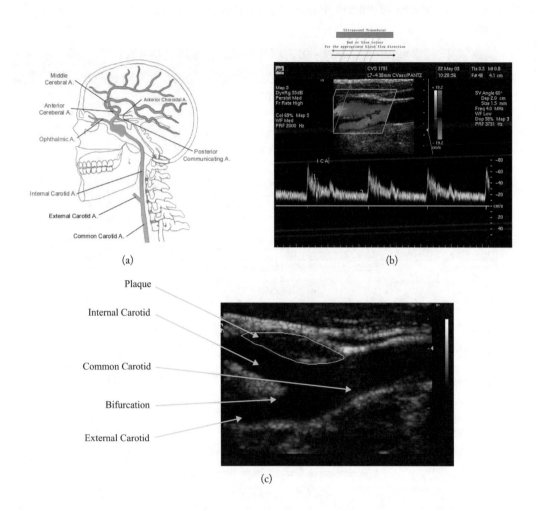

Figure 1.3: (a) The carotid system illustrating the common carotid artery, its bifurcation, and the internal and external carotid arteries (figure based on [111]); (b) longitudinal color flow duplex image of the carotid artery combined with Doppler ultrasound image. Highlighted image with white contour on top shows the carotid bifurcation. The 2D signal shows the velocity variation related to the cardiac cycle. Blood flow velocity spectrum is displayed with markings 1 and 2, where marking 1 represents the peak systolic velocity and marking 2 represents the end diastolic velocity. This is the duration of one cardiac cycle. Different colors (shades) represent blood flow direction. For the current picture, red represents the blood moving to the brain through the carotid artery, whereas blue represents the blood returning back from the brain; and (c) ultrasound B-mode longitudinal image of the carotid bifurcation with manually outlined plaque, which is usually confirmed with blood flow image.

Table 1.1: An overview of different human organs where ultrasound and other imaging modalities are applied

HUMAN ORGAN	PHYSIOLOGICAL SYSTEM	IMAGING MODALITY
Cardiovascular System		
Carotid/Aorta arteries	Stenosis, plaque, stiffness	US, MRI, IVUS
Inferior vene cava	Intravascular volume status, estimation of central venous pressure, sepsis, hydration, heart failure, haemorrhage	US, IVUS
Endocrine System		
Thyroid	Cysts, enlargement, thyroid nodules	US, CT
Pancreas	Cancer, pancreatitis	US
Muscular System		
Spine	Spine related musculoskeletal conditions	US, MRI, CT
Digestive System		
Stomach	Cancer, ulcers	US, MRI, CT
Liver	Chirosis, statosis, liver stones, liver enlargement	US, MRI, CT
Gall bladder	Stones	US, CT
Intestines	Inflammatory bowel disorders, Crohn's disease, appendicitis, diverticulitis.	US
Urinary System		
Kidneys	Stones	US, X-ray
Lymphatic System		
Spleen	Splenic rupture, haemorrhage, hematoma	US
Other		
Breast	Fibroadenoma, lumps, cysts, cancer, fibrocysts	US, MRI
Embryo/Fetus	Growth, position, development of embryo	US

US: Ultrasound, MRI: Magnetic Resonance Imaging, CT: Computed Tomography, IVUS: Intravascular Ultrasound

1.1.1 BASIC PRINCIPLES OF ULTRASOUND IMAGING AND VIDEO

Ultrasound is a sound wave with frequency that exceeds 20 KHz. It transports energy and propagates through several means as a pulsating pressure wave. It is described by a number of wave parameters such as pressure density, propagation direction, and particle displacement. If the particle displacement is parallel to the propagation direction then the wave is called longitudinal or a compression wave. If the particle displacement is perpendicular to the propagation direction,

Table 1.2: An overview of different imaging modalities degraded by speckle and other types of noise

IMAGING MODALITIES	TYPE OF NOISE
Ultrasound	Speckle, Gaussian
Intravascular Ultrasound (IVUS) Imaging	Speckle, Gaussian
Optical Coherence tomography (OCT)	Speckle, Gaussian, white noise
Optical Microscopy	Speckle, shot, dark, red
Synthetic Aperture Radar (SAR) imaging	Speckle, Gaussian, thermal, electronic

it is a shear or transverse wave. Interaction of ultrasound waves with tissue is subject to the laws of geometrical optics. It includes reflection, refraction, scattering, diffraction, interference, and absorption. Except from interference all other interactions reduce the intensity of the ultrasound beam.

The main characteristic of an ultrasound wave is its wavelength λ, which is a measure of the distance between two adjacent maximum or minimum values of a sine curve and its frequency f, which is the number of waves per unit of time. The product of these two measures gives the velocity of ultrasound wave propagation, v, described with the equation $v = f\lambda$. Ultrasound techniques are mainly based on measuring the echoes transmitted back from a medium when sending an ultrasound wave to it. In the echo impulse ultrasound technique, the ultrasound wave interacts with tissue and blood, and some of the transmitted energy returns to the transducer to be detected by the instrument. If we know the velocity of propagation in the tissue under investigation, we can determine the distance from the transducer at which the interaction occurs [20]. The characteristics of the return signal (amplitude, phase, etc.) provide information on the nature of the interaction, and hence they give some indication of the type of the medium in which they occurred. Mainly two principles are used in medical ultrasound diagnostics, the echo impulse technique and the Doppler technique [20].

The second principle used in ultrasound diagnostics is the Doppler principle, named after the physicist Christian Doppler (1803–1853) [21]. This technique is based on the principle that the received frequency of sound echoes reflected by a moving target is related to the velocity of the target. The frequency shift (Doppler frequency shift) Δf, of the echo signal is proportional to the flow velocity v cm/s, and the ultrasound transmission frequency f MHz. The Doppler shift is described by the formula $\Delta f = 2 f_0(v \cos \theta)/u_{sp}$, where f_0, is the transmitted frequency of the signal, θ, is the angle between the direction of movement of the moving object and the ultrasound beam and u_{sp}, is the speed of sound through tissue that is approximately 1540 m/s.

In Doppler ultrasound, waves are produced by a vibrating crystal using the piezoelectric effect, whereas the returned echoes are displayed as a 2D signal, as shown in Fig. 1.3b. When blood flow in a vessel is examined, sound reflections caused by the blood's corpuscular elements play a major role. Based on the fact that blood flow velocity varies in different areas of a vessel, the Doppler signal contains a broad frequency spectrum. In normal internal carotid artery (ICA) the spectrum varies from 0.5 KHz to 3.5 KHz and v is less than 120 cm/s when an ultrasound beam of 4 MHz is used.

1.1.2   ULTRASOUND MODES

The two main scanning modes are A- and B-mode. Other modes used are the M-mode, Duplex ultrasound, color coded ultrasound, and power Doppler ultrasound, which will be briefly introduced below.

A-mode refers to amplitude mode scanning, which is mainly of historical interest. In this mode the strength of the detected echo signal is measured and displayed as a continuous signal in one direction. A-mode is a line, with strong reflections being represented as an increase of signal amplitude. This scanning technique has the limitation that the recorded signal is 1D with limited anatomical information. A-mode is no longer used, especially for the assessment of cardiovascular disease. Its use is restricted to very special uses such as in ophthalmology in order to perform very accurate measurements of distance.

B-mode refers to brightness mode. In B-mode echoes are displayed as a 2D grayscale image. The amplitude of the returning echoes is represented as dots (pixels) of an image with different gray values as illustrated in Figs. 1.3b and c. The image is constructed by these pixels line by line. Advances in B-mode ultrasound have resulted in improved anatomic definition, which has enabled plaque characterization [20, 22].

The M-mode is used in cardiology and it is actually an A-scan plotted against time. The result is the display of consecutive lines plotted against time. Using this mode, detailed information may be obtained about various cardiac dimensions and also the accurate timing of vascular motion.

Moving blood (see Fig. 1.3b) generates a Doppler frequency shift in the reflected sound from insonated red blood cells and this frequency shift can be used to calculate the velocity of the moving blood, using the Doppler equation [21]. The invention of gated Doppler ultrasound in the late 1950s allowed velocity sampling at different depths and positions and its subsequent combination with B-mode real-time ultrasonic imaging led to the development of Duplex ultrasound. Stenosis in any vessel is characterized by an increase in systolic and diastolic velocities. Several types of Doppler systems are used in medical diagnosis, Continuous Wave (CW) Doppler, Pulsed Wave (PW) Doppler, Duplex ultrasound, and Color Flow Duplex (see also Fig. 1.3b). In CW Doppler, the machine uses two piezoelectric elements serving as transmitters and receivers. They transmit ultrasound beams continuously. Because of the continuous way that ultrasound is being transmitted, no specific information about depth can be obtained. PW Doppler is used in order to detect blood flow at a specific depth. Sequences of pulses are transmitted to the human body that are gated for a short period of time in order to receive the echoes. By selecting the time interval between the transmitted and received pulses, it is possible to examine vessels at a specific depth.

In color-coded ultrasound, every pixel is tested for Doppler shift. Using this technique, the movement of the red blood cells is finally depicted through color. The final image results by superimposing the color-coded image on the B-mode image.

Power Doppler is the depiction of flow, based on the integrated power of the Doppler spectrum rather than on the mean Doppler frequency. This modality results in an angle, which is independent of the resulting enhanced sensitivity in flow detection as compared to the color-coded Doppler and therefore the detection of low flow is better viewed.

Intravascular ultrasound (IVUS) is a medical imaging invasive methodology using a specially designed catheter in order to be able to acquire images from inside blood vessels [23]. The arteries of the heart and the common carotid arteries are the most frequent target of IVUS. It is used in the CCA, to determine the amount of atheromatous plaque built-up at any particular point, and to determine both plaque volume within the wall of the artery and/or the degree of stenosis of the artery lumen http://en.wikipedia.org/wiki/Lumen_(anatomy). It is especially useful in situations where angiographic imaging is unreliable. It is also used to assess the artery stenosis. A limitation of IVUS is that the 3D orientation of the transducer in the vessel is not available. Therefore, IVUS is complementary rather than in completion with conventional ultrasound but the principles of image development are the same as ultrasound. Intravascular optical coherence tomography (IVOCT) recently emerged as a novel imaging modality with the unique resolution of 10 μ m to identify vulnerable plaque characteristics [24]. This is another modality that is affected by speckle noise.

1.1.3   IMAGE AND VIDEO QUALITY AND RESOLUTION

The quality of the produced ultrasound image or video depends on image resolution, axial and lateral. Resolution is defined as the smallest distance between two points at which they can be represented as distinct. Axial resolution refers to the ability of representing two points that lie along the direction of ultrasound propagation. It depends on the wavelength of the beam. In B-mode ultrasound pulses consist of one to two sinusoidal wavelengths, and the axial resolution is dependent on the wavelength of the waveforms, and lies in the range of the ultrasound wavelength, λ ($\lambda = 0.21$ mm). Resolution depends on the frequency of the beam waveforms. Since this value is reciprocal to the ultrasound frequency ($\lambda = v/f$), the axial resolution improves with increasing frequency.

Lateral resolution refers to the ability to represent two points that lie at right angle to the direction of ultrasound propagation. This is dependent on the width of the ultrasound wave (beam). To be able to resolve points that lie close together, the width of the ultrasound beam has to be kept reasonably small and the diameter of the transducer is kept as large as possible (i.e., small phase-array transducers have a worse lateral resolution than large linear or curved-array transducers).

In order to achieve the best results in vascular ultrasound imaging, the transmission frequencies are in the range of 1–10 MHz. The selected frequency depends on the application domain. For arteries located close to the human skin, frequencies greater than 7.5 MHz are used, whereas for arteries located deeper in the human body, frequencies from 3–5 MHz are used. For transcranial applications frequencies less than 2 MHz are used. Although when selecting a frequency, the user has to keep in mind that axial resolution is proportional to the ultrasound wavelength; while the intensity of the signal depends on the attenuation of the signal transmitted through the body, with the higher the frequency the higher the attenuation. Therefore, there is a trade-off between higher resolution ultrasound images at smaller depth and lower resolution images at higher depths.

1.1.4   LIMITATIONS OF ULTRASOUND IMAGING AND VIDEO

Variability in B-mode images (even when using the same ultrasonic equipment with fixed settings) does exist [4]. Sources of variability are outlined below.

(a) Geometrical and diffraction effects, where spatial compound imaging may be employed to correct the image [21, 22].

(b) Inter-patient variation due to depth dependence and inhomogeneous intervening tissue, where normalization techniques may be applied to standardize the image [8, 9, 25, 26].

(c) Speckle noise affecting the quality of ultrasound B-mode imaging. It is described as an ultrasound textural pattern that varies depending on the type of biological tissue. The presence of speckle, which is difficult to suppress [1]–[30] may obscure small structures thus degrading the spatial resolution of an ultrasonic image [8]. Despeckle filtering may be applied to improve the quality of the image.

(d) Low contrast of the intima media complex or plaque borders [3, 7], and a small thin size [9, 25] making the image interpretation a difficult task.

(e) Falsely low echogenicity due to shadowing effects, hindering the observation in B-mode images, of plaques or the intima media complex or other structures [7].

(f) Low signal-to-noise ratio in anechoic components and difficulty in outlining the carotid plaque, or other tissue under investigation, where the difficulty may be overcome by employing the use of color-coded images [25, 26].

(g) Intra-observer variability where the ultrasound images inspected by the same expert at different occasions might be evaluated differently [8, 9].

(h) Inter-observer variability where the ultrasound images inspected by two or more experts might be evaluated differently [8].

(i) Significant processing time in case of video analysis, where optimization methods for improving the performance of the automated algorithms for video despeckling and segmentation may be employed [14, 31].

It is noted that the entries (g) and (h) are applicable in any medical imaging modality. In order to overcome intra- and inter-observer variability, it is generally recommended that multiple observers should perform the image evaluation.

1.2   SPECKLE NOISE

In this section we introduce speckle noise as a major factor limiting the visual perception and processing of ultrasound (and SAR) images and videos [13], [32]–[37]. A similar approach can also be used to IVUS and OCT images and videos which are in the scope of this book not treated. A mathematical speckle model for ultrasound images (or videos) is introduced, where the statistics of speckle noise are presented, taking into consideration the log-compression of the ultrasound image, which is performed in order to match the image into the display device (see Section 1.2.2). Based on this speckle model, a number of despeckling techniques are derived and explained in detail in Chapter 2. The same speckle model can also be used for ultrasound video despeckling. Specifically, the following categories of despeckle filtering techniques are presented: linear filtering (local statistics filtering, homogeneity filtering), nonlinear filtering (median filtering, linear scaling filtering, geometric filtering, logarithmic filtering, homomorphic filtering), anisotropic diffusion filtering (anisotropic diffusion, speckle reducing anisotropic diffusion, coherent nonlinear anisotropic diffusion), and wavelet filtering.

Noise and artefacts can cause signal, image, and video degradations for many medical image and video modalities. Different image modalities exhibit distinct types of degradation. Images formed with coherent energy, such as ultrasound, suffer from speckle noise. Image degradation can have a significant impact on image quality and thus affect human interpretation and the accuracy of computer-assisted methods. Poor image/video quality often makes feature extraction, analysis, recognition, and quantitative measurements problematic and unreliable. Therefore, despeckling is a very important task, which motivated a significant number of studies in medical imaging [3, 5, 13, 27, 28, 38–40] and video [14, 15, 31, 37].

The use of ultrasound in the diagnosis and assessment of arterial disease is well established because of its non-invasive nature, its low cost, and the continuing improvements in image quality [1]. Speckle, a form of locally correlated multiplicative noise that corrupts medical ultrasound imaging and video making visual observation difficult [32, 33]. The presence of speckle noise in ultrasound images has been documented since the early 1970s where researchers such as Burckhardt [32], Wagner [33], and Goodman [34], described the fundamentals and the statistical properties of the speckle noise. Speckle is not truly a noise in the typical engineering sense, since its texture often carries useful information about the image being viewed [32–34].

Speckle noise is the primary factor which limits the contrast resolution in diagnostic ultrasound imaging, thereby limiting the detectability of small, low-contrast lesions and making the ultrasound images generally difficult for the non-specialist to interpret [29, 32, 33, 35]. Due to the speckle presence, ultrasound experts with sufficient experience may not often draw useful conclusions from the images [35]. Speckle also limits the effective application (e.g., edge detection) of automated computer aided analysis (e.g., volume rendering and 3D display) algorithms. It is caused by the interference between ultrasound waves reflected from microscopic scattering through the tissue.

Therefore, speckle is most often considered a dominant source of noise in ultrasound imaging and should be filtered out [29, 32, 35] without affecting important features of the image. In this book we carry out a comparative evaluation of despeckle filtering techniques based on texture analysis, image and video quality evaluation metrics as well as visual assessment by experts on 440 ultrasound images and 80 ultrasound videos of the carotid artery bifurcation. Results of this study were also published in [7, 13, 15]. Moreover, a comparative evaluation framework for the selection of the most appropriate despeckle filter for the problem under investigation is proposed.

1.2.1 PHYSICAL PROPERTIES AND PATTERN OF SPECKLE NOISE

The speckle pattern, that is visible as the typical light and dark spots the image is composed of, results from destructive interference of ultrasound waves scattered from different sites. The nature of speckle has been a major subject of investigation [32–34, 40, 41]. When a fixed, rigid object is scanned twice under exactly the same conditions, one obtains identical speckle patterns. Although of random appearance, speckle is not random in the same sense as electrical noise. However, if the same object is scanned under slightly different conditions, say with a different transducer aperture, pulse length or transducer angulation, the speckle patterns change.

The most popular model adopted in the literature to explain the effects that occur when a tissue is insonated is illustrated in Fig. 1.4, where a tissue may be modeled as a sound absorbing medium containing scatterers, which scatter the sound waves [42, 43]. These scatterers arise from inhomogeneities and structures approximately equal to or smaller in size than the wavelength of the ultrasound, such as tissue parenchyma, where there are changes in acoustic impedance over a microscopic level within the tissue. Tissue particles that are relatively small in relation to the wavelength (i.e., blood cells), and particles with differing impedance that lie very close to one another, cause scattering or speckling. Absorption of ultrasound tissue is an additional factor to scattering and refraction, responsible for pulse energy loss. The process of energy loss involving absorption, reflection, and scattering is referred to as attenuation, which increases with depth and frequency. Because higher frequency of ultrasound, results into increased absorption, the consequence is a decrease of the depth of visualisation.

Figure 1.5 illustrates the entire scattering procedure [42]. Consider a transducer insonating a homogeneous medium containing four point-like scatterers, as depicted in Fig. 1.5a. These scatterers yield spherical waves that will arrive at the transducer at slightly different times after the transmission of the ultrasound pulse. Usually, the pulse envelope is approximately Gaussian as shown in Fig. 1.5b. If the pulse has a Gaussian shape then so has its spectrum. One chooses a Gaussian shape, because for a medium with linear attenuation coefficient this Gaussian shape of the spectrum is maintained while the pulse travels through the medium (although a shift of

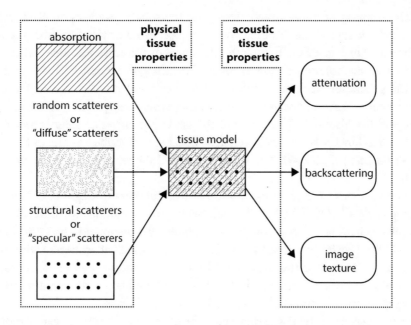

Figure 1.4: The usual tissue model in ultrasound imaging, modified from [42].

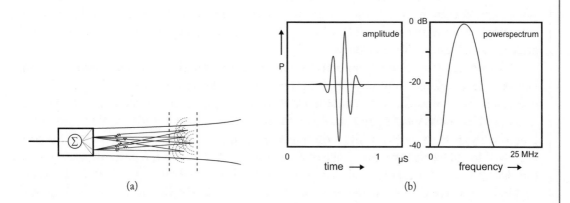

Figure 1.5: (a) The scattering in the sound beam and (b) one pulse in the time and frequency domains (from [42]).

this Gaussian spectrum to lower frequencies occurs while the pulse travels through the medium, because the attenuation increases with the frequency).

Upon reception of the reflected signal, the transducer produces an electrical signal (RF) that is the algebraic sum of the instantaneous sound pressures originating from the backscattered waves (four waves in Fig. 1.5a). The depth differences of the scatterers are smaller than the axial size of the resolution volume of the transducer (i.e., the pulse length). This is, in fact, the basic cause for the generation of tissue texture. The formed pattern is the so-called speckle pattern. Note, in particular, that the tissue texture resulting from this speckle pattern is in general not a true image of the histological structure of the tissue but rather an interference pattern that is mainly determined by the beam characteristics. Speckle is described as one of the more complex image noise models [33, 34, 40, 42, 44] it is a signal dependent, non-Gaussian and spatially dependent.

In homogeneous tissue, the distribution of the scatterers throughout 3D space is assumed to be isotropic. As displayed in Fig. 1.4 one distinguishes random (or diffuse) scatterers, and structural (or specular) scatterers. The diffuse scatterers are assumed to be uniformly distributed over space. Diffuse scattering arises when there are a number of scatterers with random phase within the resolution cell of the ultrasound beam. This random nature of the location of the scatterers causes the statistical nature of the echo signals, and hence the resulting speckle pattern. Consequently, a statistical approach to its analysis seems obvious.

Other properties of the tissue that affect the ultrasound as it propagates through it are the propagation speed, the attenuation, and the backscattering. The absorption of ultrasound is caused by relaxation phenomena of biological macromolecules [46] that transfer mechanical energy into heat. Another source of attenuation is the scattering, i.e., omni-directional reflections by small inhomogeneities in the tissue. The overall attenuation is therefore the resultant of absorption and scattering (as illustrated in Fig. 1.4), which are both frequency-dependent in such a way that the attenuation increases with frequency.

In analyzing speckle an important point to bear in mind is to make a clear distinction between the speckle as it appears in the image and the speckle in the received RF-signal. The block diagram in Fig. 1.6 explains the entire track of the RF-signal from the transducer to the screen inside the ultrasound imaging system. As set forth, the signal is subject to several transformations that severely affect its statistics. The most important of these is the log-compression of the signal, employed to reduce the dynamic range of the input signal to match the lower dynamic range of the display device. The input signal could have a dynamic range of the order of 50–70 dB whereas a typical display could have a dynamic range of the order of 20–30 dB. Such a relation is normally affected through an amplifier, which has a reducing amplification for a larger input signal.

In addition, the expert has the possibility to adjust several machine settings manually. In Fig. 1.6 these are indicated as the slide contacts overall gain and time gain compensation (TGC). These machine settings control the amplification of the signal, the overall gain controls the overall amplification, and the TGC is a time-dependent amplification, and serve as tools for the expert to

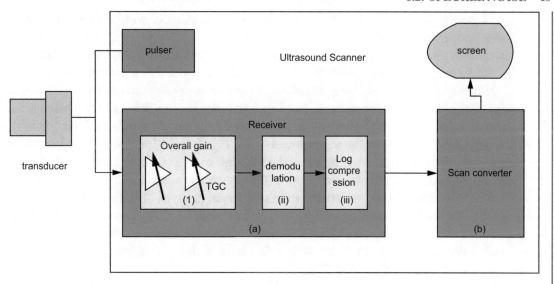

Figure 1.6: The processing steps of the RF-signal inside the ultrasound scanner, modified from [45].

adjust the image for an optimal visual diagnosis. The TGC is adjusted by several (usually seven) slide contacts, each of which controls the gain in part of the image. For instance, if the slide contacts are placed in a vertical row, the top slide contact controls the gain in the top of the image, the bottom slide contact controls the gain in the bottom of the image, etc. This position-specific gain in the image is realized by making the amplification of the signal dependent on the exact time that the sound reflection is received. Since the position where a pixel is displayed on the screen is dependent on this time instant, the time-dependent amplification of the received signal converts to a position dependent change in gray value of the pixels on the screen.

1.2.2 SPECKLE NOISE MODELING

In order to be able to derive an efficient despeckle filter, a speckle noise model is needed. The speckle noise model for both ultrasound and SAR images and videos may be approximated as multiplicative [40]. The signal at the output of the receiver demodulation module of the ultrasound imaging system (see Fig. 1.6a (ii)), may be defined as:

$$y_{i,j} = x_{i,j} n_{i,j} + a_{i,j}, \tag{1.1}$$

where $y_{i,j}$ represents the noisy pixel in the middle of the moving window, $x_{i,j}$ represents the noise-free pixel, $n_{i,j}$ and $a_{i,j}$ represent the multiplicative and additive noise, and i, j are the indices of the spatial locations that belong in the 2D space of real numbers, $i, j \in \Re^2$.

Despeckling is based on estimating the true intensity $x_{i,j}$, as a function of the intensity of the pixel $y_{i,j}$, and some local statistics calculated on a neighborhood of this pixel.

Wagner et al. [33] showed that the histogram of amplitudes within the resolution cells of the envelope-detected RF-signal backscattered from a uniform area with a sufficiently high scatterer density has a Rayleigh distribution with mean (μ) proportional to the standard deviation (σ), (with $\mu/\sigma = 1.91$). This implies that speckle could be modeled as a multiplicative noise.

However, the signal processing stages inside the scanner modify the statistics of the original signal, i.e., the logarithmic compression (see Fig. 1.6a (iii)). The logarithmic compression is used in order to adjust the large echo dynamic range (50–70 dB) to the number of bits (usually 8) of the digitization module in the scan converter (see Fig. 1.6b). More specifically, logarithmic compression affects the high intensity tail of the Rayleigh and Rician probability density function (PDF) more than the low intensity part. As a result, the speckle noise becomes very close to white Gaussian noise corresponding to the uncompressed Rayleigh signal [40]. In particular, it should be noted, that speckle is no longer multiplicative in the sense that on homogeneous regions, where $x_{i,j}$, can be assumed constant, the mean is proportional to the variance ($\mu \approx \sigma^2$), rather than the standard deviation ($\mu \approx \sigma$), [3, 5, 28, 40]. In this respect, the speckle index, C, will be for the log-compressed ultrasound images, $C = \sigma^2/\mu$.

Referring back to Eq. (1.1), since the effect of additive noise is considerably smaller compared with that of multiplicative noise, it may be written as:

$$y_{i,j} \approx x_{i,j} n_{i,j}. \tag{1.2}$$

Thus, the logarithmic compression transforms the model in (1.2) into the classical signal in additive noise form as:

$$\log(y_{i,j}) = \log(x_{i,j}) + \log(n_{i,j}), \tag{1.3}$$

and

$$g_{i,j} = f_{i,j} + nl_{i,j}. \tag{1.4}$$

For the rest of the book, the term $\log(y_{i,j})$, which is the observed pixel on the ultrasound image display after logarithmic compression, is denoted as $g_{i,j}$, and the terms $\log(x_{i,j})$, and $\log(n_{i,j})$ which are the noise free pixel and noise component after logarithmic compression, as $f_{i,j}$ and $nl_{i,j}$, respectively (see Eq. (1.4)).

1.2.3   EARLY ATTEMPTS OF DESPECKLE FILTERING IN DIFFERENT MODALITIES AND ULTRASOUND IMAGING AND VIDEO

The widespread of ultrasound imaging equipment, including mobile and portable telemedicine ultrasound scanning instruments and computer-aided systems, necessitate the need for better image processing techniques, in order to offer a clearer image to the medical practitioner. This makes the use of efficient despeckle filtering a very important task. Early attempts to suppress speckle noise were implemented by averaging of uncorrelated images of the same tissue recorded under different spatial positions [29, 36, 44]. While these methods are effective for speckle reduction, they require multiple images of the same object to be obtained [46]. Speckle reducing filters originated from the synthetic aperture radar (SAR) community [36]. These filters have then later been applied to ultrasound imaging since the early 1980s [41]. Filters that are used widely in both SAR and ultrasound imaging were proposed originally by Frost [47], Lee [36, 38, 48], and Kuan [47, 49, 50].

Some researchers have tried in the past to despeckle SAR images by averaging of uncorrelated images obtained from different spatial positions [19]. These temporal averaging and multi-frame methods aimed to increase the SNR by generating multiple uncorrelated images that are summed incoherently to reduce speckle [51]. Despite being simple and fast, these approaches suffer from two limitations. First, in order to produce uncorrelated ultrasound images, the transducer has to be translated at least by about half its element width for each of the generated frames [32]. Second, temporal averaging based on transducer movement causes the loss of small details such as small vessels and texture patterns because of blurring. For the above reasons this procedure has been proven to be not suitable for despeckle filtering. It is most suitable for additive noise reduction [19, 51]. Another disadvantage of this method is that multiple images from the same object are required [44, 48]. Other researchers applied their techniques on ultrasound images of the kidney [5], echocardiograms [52], heart [3], abdomen [3], pig heart [28], liver [53], SAR images [51, 54–56], real world [28, 50], and artificial images [44, 57]. They used statistical measures like the mean, variance, median, the speckle index (C), the mean-square error (MSE), image contrast, and visual perception evaluation made by experts, to evaluate their techniques. They compared their despeckling techniques with the Lee filter [36], homomorphic filtering [51, 58, 59], media filter [60], and diffusion filtering [27, 29, 39, 61, 62]. Despeckle filtering can also be used as a pre-processing step for image segmentation [7–9, 25], image registration [19], or techniques. By suppressing the speckle the performance of these techniques can be improved.

Many authors have shown a reduction of lesion detectability of approximately a factor of eight due to the presence of speckle noise in the image [32, 34, 47]. This radical reduction in contrast resolution is responsible for the poorer effective resolution of ultrasound compared to X-ray and MRI [19]. Despeckle filtering is therefore a critical pre-processing step in medical ultrasound images, provided that the features of interest for diagnosis are not lost.

1.2.4   SPECKLE NOISE TRACKING

Ultrasound speckle noise tracking is a technique that allows the assessment of tissue motion and deformation by tracking interference patterns across imaging video frames [63]. The technique has to a large extent been developed and applied for the assessment of the mechanical properties of the myocardium [64], whereas speckle tracking based carotid arterial strain assessment has gained interest in recent years [63–65]. Speckle tracking and tissue Doppler imaging techniques have shown potential in both subclinical detection of increased arterial stiffness, as lower arterial strain values were associated with increased cardiovascular risk [66], and in the assessment of carotid plaque characteristics to predict plaque rupture, as strain correlated with plaque composition [63]. Assessing strain in the arterial wall and in atherosclerotic plaques is particularly challenging because of the small structures involved and their low physiologic deformation in relation to the applied ultrasound wavelength used in clinical ultrasound systems. A variety of methods enabling measurements of radial and circumferential arterial strain have been developed and applied both in phantom setups and in vivo [63–66]. During the last decade, methods to assess motion and strain in the longitudinal axis of the arterial wall have also been presented [67, 68]. The longitudinal motion of the artery has been neglected because it is difficult to assess due to the low amplitudes combined with the intrinsic lower spatial resolution in the azimuth direction.

1.3 AN OVERVIEW OF DESPECKLE FILTERING TECHNIQUES

Table 1.3 summarizes the despeckle filtering techniques for ultrasound imaging that are presented in this book, grouped under the following categories: linear filtering (local statistics filtering, homogeneity filtering), nonlinear filtering (median filtering, linear scaling filtering, geometric filtering, logarithmic filtering, homomorphic filtering, hybrid median filtering, Kuwahara filtering), anisotropic diffusion filtering (anisotropic diffusion, speckle reducing anisotropic diffusion, coherent nonlinear anisotropic diffusion), and wavelet filtering. Furthermore, in Table 1.3 the methodology used, the main investigators, and the corresponding filter names are given. These filters are briefly introduced in this chapter, and presented in detail in Chapters 3–6.

Some of the linear filters are the Lee [36, 38, 48], Frost [47], and Kuan [46, 50]. The Lee and Kuan filters have the same structure, whereas the Kuan is a generalization of the Lee filter. Both filters form the output image by computing the central pixel intensity inside a filter-moving window, which is calculated from the average intensity values of the pixels and a coefficient of variation inside the moving window. Kuan considered a multiplicative speckle model and designed a linear filter, based on the minimum-mean-square error (MMSE) criterion that has optimal performance when the histogram of the image intensity is Gaussian distributed. The Lee [36] filter is a particular case of the Kuan filter based on a linear approximation made for the multiplicative noise model. The Frost [47] makes a balance between the averaging and the all-pass filters. It was designed as an adaptive *DsFwiener* filter that assumed an autoregressive exponential model for the image.

In the nonlinear filtering group the gray level values are linearly scaled to despeckle the image [69]. Some of the nonlinear filters are based on the most homogeneous neighborhood around each image pixel [2]. Geometric filters [44] are based on nonlinear iterative algorithms, which increment or decrement the pixel values in a neighborhood based upon their relative values. The method of homomorphic filtering [51, 59] is similar to the logarithmic point operations used in histogram enhancement, where dominant bright pixels are de-emphasized. In the homomorphic filtering, the FFT of the image is calculated then denoised, and then the inverse FFT is calculated. Finally, hybrid median (*DsFhmedian*) filtering and Kuwahara (*DsFKuwahara*) filtering are special cases of the median filtering (*DsFmedian*).

Table 1.3: An overview of despeckle filtering techniques

Speckle Reduction Technique	Method	Investigator	Filter Name
Linear filtering	Moving window utilising local statistics		
	(a)mean (m), variance (σ^2)	[36–38],[47–50]	*DsFlsmv*
	(b)mean, variance, 3rd and 4th moments (higher statistical moments) and entropy	[36–38]	*DsFlsmvsk1d* *DsFlsmvsk2d*
	(c)homogeneous mask area filters	[79, 80]	*DsFlsminsc*
	(d)Wiener filtering	[32–38, 48]	*DsFwiener*
Nonlinear Filtering	Median filtering	[60]	*DsFmedian*
	Linear scaling of the gray level values	[19]	*DsFls* *DsFca* *DsFlecasort*
	Based on the most homogeneous neighborhood around each pixel	[2]	*DsFhomog*
	Nonlinear iterative algorithm (geometric filtering)	[44]	*DsFgf4d*
	The image is logarithmically transformed, the Fast Fourier transform (FFT) is computed, denoized, the inverse FFT is computed and finally exponentially transformed back	[32, 51, 59]	*DsFhomo*
	Hybrid median filtering	[81]	*DsFhmedian*
	Kuwahara filtering	[82]	*DsFKuwahara*
	Nonlocal filtering	[83]	*DsFnlocal*

Diffusion filtering	Nonlinear filtering technique for simultaneously performing contrast enhancement and noise reduction	[27, 29, 32, 41] [47, 60, 61, 85]	*DsFad*
	Exponential damp kernel filters utilizing diffusion	[29]	
	Speckle reducing anisotropic diffusion based on the coefficient of variation	[3]	*DsFsrad*
	Nonlinear anisotropic diffusion	[3]	*DsFnldif*
	Nonlinear complex diffusion	[69]	*DsFncdf*
Wavelet filtering	Threshold wavelet coefficients based on speckle noise at different levels	[28, 50, 70, 71] [84, 86]	*DsFwaveltc*

The diffusion filtering category, includes filters based on anisotropic diffusion [27, 32, 60, 61], coherent anisotropic diffusion [3], speckle reducing anisotropic diffusion [29], and nonlinear complex diffusion [69]. These filters have been recently presented in the literature, and are nonlinear filtering techniques. They simultaneously perform contrast enhancement and noise reduction by utilising the coefficient of variation [29]. Furthermore, in the wavelet category, filters for suppressing the speckle noise were documented. These filters are making use of a realistic distribution of the wavelet coefficients [32, 50, 60, 70] where only the useful wavelet coefficients are utilized. Different wavelet shrinkage approaches were investigated extensively based on Donoho's work [71].

While there are a number of techniques for ultrasound image despeckling proposed in the literature, we found [15] no other studies in the literature for despeckle filtering in ultrasound videos of the CCA. There are, however, several other studies reported in the literature for filtering additive noise from natural video sequences [72–77], but we have found no other studies where despeckle filtering on ultrasound medical videos (of the CCA) was investigated. Previous research on the use of despeckle filtering of the CCA images was also reported by our group in [7, 8, 12–15, 37, 51, 78], where improved results were presented in terms of visual quality and classification accuracy between asymptomatic and symptomatic plaques. Moreover, it should be mentioned that a significant number of studies investigated different despeckle filters in various medical ultrasound video modalities with very promising results [78]. The usefulness of these methods in ultrasound video denoising on multiplicative noise still remains to be investigated.

Figure 1.7 illustrates an original longitudinal asymptomatic (see Fig. 1.7a) and symptomatic image (see Fig. 1.7e) and their despeckled images (see Fig. 1.7b and Fig. 1.7f), respectively. Asymptomatic images were recorded from patients at risk of atherosclerosis in the absence of clinical symptoms, whereas symptomatic images were recorded from patients at risk of atherosclero-

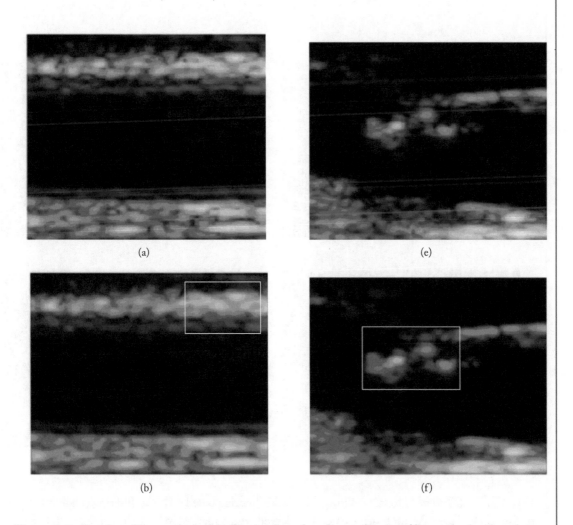

Figure 1.7: Results of image despeckle filtering based on linear filtering (first-order local statistics, *DsFlsmv*). Asymptomatic case: (a) original, (b) despeckled. Symptomatic case: (e) original, (f) despeckled, (e) enlarged region marked in (f) of the original. Regions were enlarged by a factor of three. *(Continues.)*

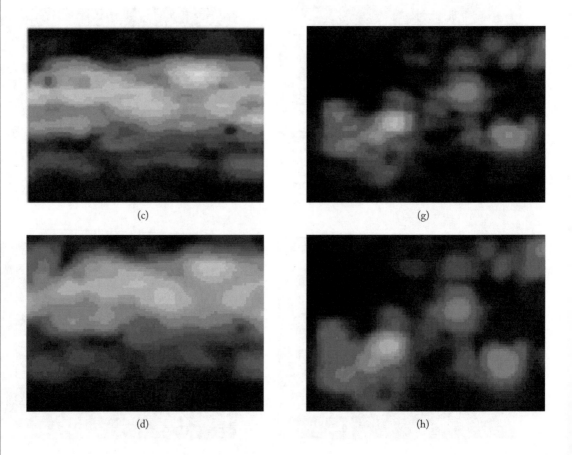

(c)   (g)

(d)   (h)

Figure 1.7: *(Continued.)* Results of image despeckle filtering based on linear filtering (first-order local statistics, *DsFlsmv*). Asymptomatic case: (c) enlarged region marked in b) of the original, (d) enlarged region marked in (b) of the despeckled image. Symptomatic case: (h) enlarged region marked in (f) of the despeckled image. Regions were enlarged by a factor of three.

sis, which have already developed clinical symptoms, such as a stroke episode. Figures 1.7c–1.7h show an enlarged window from the original and despeckled images (shown in a rectangle in Fig. 1.7b and Fig. 1.7f).

Figure 1.8 presents the 100th frame of an ultrasound video of the CCA from a symptomatic subject for the original and the despeckled frames with filters *DsFlsmv*, *DsFhmedian*, *DsFkuwahara*, and *DsFsrad* when applied to the whole video frame (left column) and on an region of interest (ROI) selected by the user of the system (right column), respectively, (see list of filters tabulated in Table 1.3). The automated plaque segmentations performed by an integrated system proposed in [14] are also shown in the images. The filters *DsFlsmv* and *DsFhmedian* smoothed the video frame without destroying subtle details.

1.4   LIMITATIONS OF DESPECKLE FILTERING TECHNIQUES

Despeckling is always a trade-off between noise suppression and loss of information, something that experts are very concerned about. It is therefore desirable to keep as much of important information as possible. The majority of speckle reduction techniques have certain limitations that can be briefly summarized as follows.

(a) They are sensitive to the size and shape of the window. The use of different window sizes greatly affects the quality of the processed images. If the window is too large over smoothing will occur, subtle details of the image will be lost in the filtering process and edges will be blurred. On the other hand, a small window will decrease the smoothing capability of the filter and will not reduce speckle noise thus making the filter not effective. In homogenous areas, the larger the window size, the more efficient is the filter in reducing the speckle noise. In heterogeneous areas the smaller the window size, the more it is possible to keep subtle image details unchanged. Our experiments showed that a 7×7 window size is a fairly good choice.

(b) Some of the despeckle methods based on window approaches require thresholds to be used in the filtering process, which have to be estimated empirically. There are a number of thresholds introduced in the literature which includes gradient thresholding [29], soft or hard thresholds [71], nonlinear thresholds [28], and wavelet thresholds [28, 70, 87, 88]. The inappropriate choice of a threshold may lead to average filtering and noisy boundaries thus leaving the sharp features unfiltered [7, 38, 44].

(c) Most of the existing despeckle filters do not enhance the edges but they only inhibit smoothing near the edges. When an edge is contained in the filtering window, the coefficient of variation will be high and smoothing will be inhibited. Therefore, speckle in the neighborhood of an edge will remain after filtering. They are not directional in the sense that in the presence of an edge, all smoothing is precluded. Instead of inhibiting smoothing in directions perpendicular to the edge, smoothing in directions parallel to the edge is allowed.

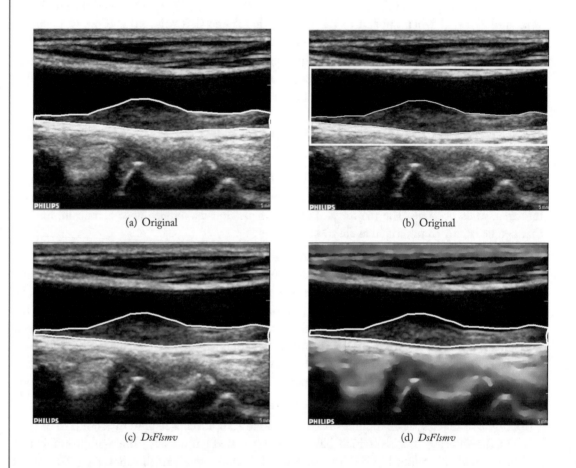

(a) Original

(b) Original

(c) *DsFlsmv*

(d) *DsFlsmv*

Figure 1.8: Examples of despeckle filtering on a video frame of a CCA video acquired from a male symptomatic subject at the age of 62 with 40% stenosis and a plaque at the far wall of the CCA, for the whole image frame in the left column, and on an ROI (including the plaque (shown in (b)), in the right column for: (a), (b) original and (c), (d) *DsFlsmv*. *(Continues.)*

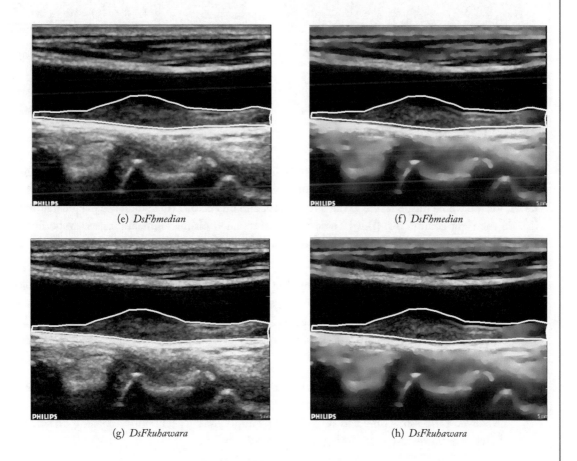

(e) *DsFhmedian* (f) *DsFhmedian*

(g) *DsFkuhawara* (h) *DsFkuhawara*

Figure 1.8: *(Continued.)* Examples of despeckle filtering on a video frame of a CCA video acquired from a male symptomatic subject at the age of 62 with 40% stenosis and a plaque at the far wall of the CCA, for the whole image frame in the left column, and on an ROI (including the plaque (shown in (b)), in the right column for: (e), (f) *DsFhmedian* and (g), (h) *DsFkuwahara*. *(Continues.)*

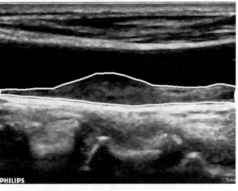 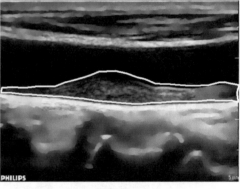

(i) *DsFsrad*   (j) *DsFsrad*

Figure 1.8: *(Continued.)* Examples of despeckle filtering on a video frame of a CCA video acquired from a male symptomatic subject at the age of 62 with 40% stenosis and a plaque at the far wall of the CCA, for the whole image frame in the left column, and on an ROI (including the plaque (shown in (b)), in the right column for: (i), (j) *DsFsrad.* The automated plaque segmentations were performed with the system proposed in [14] and are shown in all examples.

(d) Different evaluation criteria for evaluating the performance of despeckle filtering are used by different studies. Although most of the studies use quantitative criteria like the mean square error (MSE) and speckle index (C), there are additional quantitative criteria, like texture analysis and classification, image quality evaluation metrics and visual assessment by experts that could be investigated.

(e) Computational efficiency can be a limiting factor. For example, running the despeckle algorithms in MATLAB™ might not be the best computationally efficient environment. However, processing time can be reduced by applying despeckle filtering only on selected ROIs as demonstrated also in Fig. 1.8 right column.

(f) Evaluation and validation of the final results are two of the most challenging tasks in medical image processing and analysis applications [13] including despeckle filtering. Results are usually compared with the optical perception evaluation by the experts. It is true that such comparisons are often affected by other issues, such as, for example, inter- and intra-observer variability [12–15, 37, 40, 71].

1.5   GUIDE TO BOOK CONTENTS

In the following chapter, the evaluation methodology based on texture and statistical analysis, image quality evaluation metrics, and the experiments carried out for visual evaluation are presented.

In Chapters 3–6, we present the theoretical background (equations), the algorithmic steps, and the MATLAB™ code of selected despeckle filters given in Table 1.3, for linear despeckle filtering, for nonlinear despeckle filtering, for diffusion despeckle filtering and for wavelet despeckle filtering. Chapter 7 discusses, compares, and evaluates the proposed despeckle filtering techniques. Chapter 8 presents the summary and future directions, where a despeckling filtering protocol is also proposed. Finally, at the end of this book, an Appendix provides details about the despeckle filtering MATLAB™ toolbox.

CHAPTER 2

Basics of Evaluation Methodology

In this chapter we present an ultrasound phantom, and the generation of artificial images and videos, which are used to evaluate despeckle filtering. Furthermore, the Image Despeckle Filtering (IDF) and Video Despeckle Filtering (VDF) software toolboxes are presented. Moreover, a number of image quality metrics are presented for evaluating the quality between the original and the despeckled images.

2.1 USE OF PHANTOM AND ARTIFICIAL ULTRASOUND IMAGES AND VIDEOS

Ultrasound phantoms are used to evaluate axial resolution, lateral resolution, depth calibration, dead zone measurement, registration with different backgrounds, and others. Moreover, Doppler flow phantoms test both Doppler and B-mode ultrasound video to evaluate the flow rate, visibility for various angles, beam directions and operating modes, determining also the similarities between B-mode and color flow images in real time [89]. Most clinicians rely on Doppler ultrasound measurements, but a subject's maximum blood velocity rates, can vary by 10–60%, even using the same machine, due to calibration problems. The same may be observed with the speckle pattern. A different speckle pattern may be obtained from the same patient using the same machine. As a result it is impossible to obtain an accurate measurement in time. In theory we can get around this problem, if we had independent information on blood flow and speckle pattern, through human arteries, against which to calibrate the (Doppler) ultrasound data. Nevertheless, in order to get truly independent and accurate information in practice is difficult and therefore the use of the phantoms is required [90]. An example of an ultrasound phantom used to evaluate despeckle filtering is illustrated in Fig. 2.1a.

To further evaluate despeckle filtering, an artificial carotid image was generated (see Fig. 2.1b). Despeckle filtering was evaluated visually by two experts (a cardiovascular surgeon and a neurovascular specialist), on the artificial carotid image corrupted by speckle noise. The artificial image (shown in Fig. 2.1b), has a resolution of 150×150 pixels, and was generated with gray level values of the bottom, strip, middle, and upper segments of 182, 250, 102, and 158 respectively. This image was corrupted by speckle noise (see Fig. 7.1a), which was generated using the equation: $g_{i,j} = f_{i,j} + n_{i,j} f_{i,j}$, where $g_{i,j}$, and $f_{i,j}$, are the noisy and the original

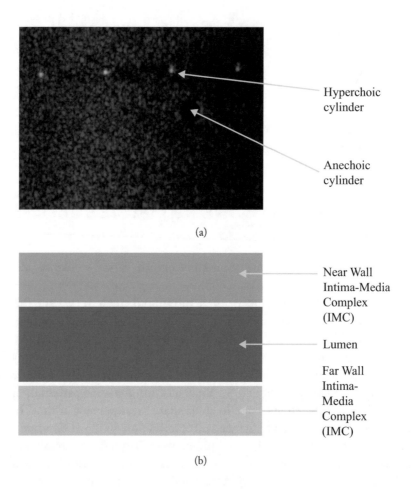

Figure 2.1: (a) Phantom ultrasound image and (b) longitudinal view of an artificial carotid ultrasound image. *(Continues.)*

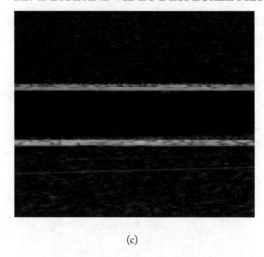

(c)

Figure 2.1: *(Continued.)* (c) Longitudinal view of an artificial carotid ultrasound frame.

images respectively, and $n_{i,j}$, a uniformly distributed random noise with mean 0 and a variance $\sigma_n^2 = 0.07$.

In order to evaluate despeckle filtering in video, an artificial carotid video was also generated (see Fig. 2.1c) [91] using the Field II–based ultrasound simulation software tool where the arterial wall motion was mathematically modeled. The video frames were also corruptred with multiplicative noise with mean 0 and a variance $\sigma_n^2 = 0.07$.

2.2 IMAGE AND VIDEO DESPECKLE FILTERING TOOLBOXES

Twenty despeckle filters were selected and investigated as presented in detail in Chapters 3–6. These filters that are also tabulated in Table 1.3 are the following: *DsFlsmv, DsFlsmvsk1d, DsFlsmvsk2d, DsFlsminsc, DsFwiener*, from the linear filtering group, *DsFmedian, DsFls, DsFca, DsFcasort, DsFhomog, DsFgf4d, DsFhomo, DsFhmedian, DsFKuwahara, DsFnlocal* from the nonlinear filtering group, *DsFad, DsFsrad, DsFnldif, DsFncdif* from the diffusion filtering group and the *DsFwaveltc* wavelet despeckle filter. These filters as well as all their corresponding functionality were integrated in the IDF software toolbox [13] (screenshots of the IDF toolbox are shown in Fig. 2.2) as well in the VDF software toolbox [37].

It should be noted that the procedure for image and video despeckling is similar. For image despeckling the filters can be applied to each image: (a) for the whole image or (b) for a region of interest selected by the user. For video despeckling, the filters can be applied to each video frame: (a) for the whole video frame or (b) for an ROI in each video frame. The number of iterations, the window size and other filtering parameters that are used for each despeckle filtering method were tuned based on the subjective despeckled image or video evaluation by the medical experts, as well as the image quality evaluation on the image and video quality metrics that are presented in the following section.

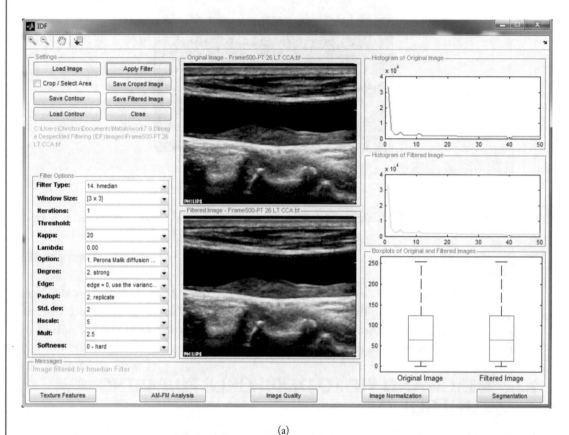

(a)

Figure 2.2: The graphical user interface of the IDF toolbox [13]: (a) filter section and application. *(Continues.)*

2.3   IMAGE AND VIDEO QUALITY EVALUATION METRICS

For medical images and videos, quality can be objectively defined in terms of performance in clinically relevant tasks such as lesion detection and classification, where typical tasks are the detection of an abnormality, the estimation of some parameters of interest, or the combination of the above [92]. Most studies today have assessed the equipment performance by testing diagnostic performance of multiple experts, which also suffer from intra- and inter-observer variability.

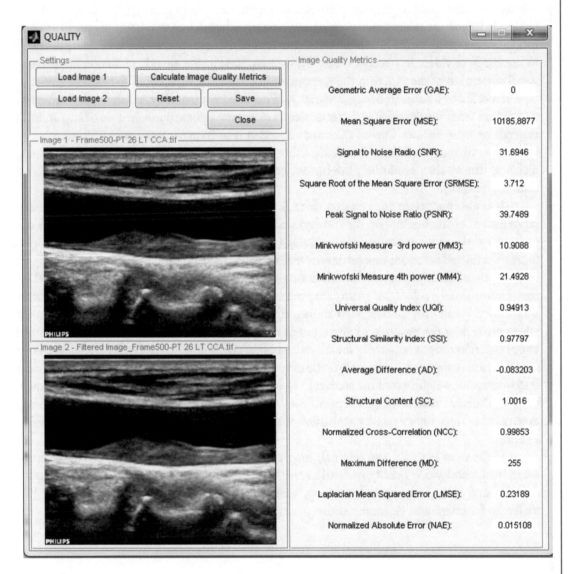

(b)

Figure 2.2: *(Continued.)* The graphical user interface of the IDF toolbox [13]: (b) computing the image quality evaluation metrics.

Although this is the most important method of assessing the results of image degradation, few studies have attempted to perform physical measurements of degradation [93]. Image and video quality is important when evaluating or segmenting atherosclerotic carotid plaques [9] or the intima-media-complex (IMC) in the carotid artery [25], where speckle obscures subtle details [7] in the image or video. In a recent study [8], we have shown that speckle reduction improves the visual perception of the expert in the assessment of ultrasound imaging of the carotid artery.

In order to be able to design accurate and reliable image quality metrics, it is necessary to understand what quality means to the expert. An expert's satisfaction when watching an image depends on many factors. One of the most important is of course image content. Research made in the area of image quality showed, that this depends on many parameters, such as: viewing distance, display size, resolution, brightness, contrast, sharpness, colorfulness, naturalness, and other factors [94].

It is also important to note that there is often a difference between fidelity (the accurate reproduction of the original on the display) and perceived quality. Sharp images with high contrast are usually more appealing to the average expert. Likewise, humans usually prefer slightly more colorful and saturated images despite realizing that they look somewhat unnatural [95]. For studying visual quality some of the definitions above should be related to the human-visual system. Unfortunately, subjective quality may not be described by an exact figure, due to its inherent subjectivity, it can only be described statistically. Even in psychological threshold experiments, where the task of the expert is to give a yes or no answer, there exists a significant variation between expert's contrast sensitivity functions and other critical low-level visual parameters. When speckle noise is apparent in the image, the expert's differing experiences with noise are bound to lead to different weightings of the artifact [94]. Researchers showed that experts and non-experts, examine different critical image characteristics to form their final opinion with respect to image quality [96]. Thus, image quality evaluation metrics can be used for the evaluation of despeckle filtering.

Differences between the original, $g_{i,j}$, and the despeckled, $f_{i,j}$, images were evaluated using image and video quality evaluation metrics. The following measures, which are easy to compute and have clear physical meaning, were computed using the IDF [13] and VDF [15], toolboxes for image and video, respectively.

(a) The *MSE*

$$MSE = \frac{1}{MN} \sum_{i=1}^{M} \sum_{j=1}^{N} (g_{i,j} - f_{i,j})^2, \qquad (2.1)$$

which measures the quality change between the original and processed image in an $M \times N$ window [97]. The *MSE*, has been widely used to quantify image quality and when is used alone, it does not correlate strongly enough with perceptual quality. It should be used therefore together with other quality metrics and visual perception [97, 98].

(b) The root *MSE* (*RMSE*), which is the square root of the squared error averaged over an $M \times N$ window [19]:

$$RMSE = \sqrt{\frac{1}{MN} \sum_{i=1}^{M} \sum_{j=1}^{N} (g_{i,j} - f_{i,j})^2}. \qquad (2.2)$$

The popularity of *RMSE* arises mostly from the fact that is in general the best approximation of the standard error.

(c) The error summation in the form of the Minkowski metric, which is the norm of the dissimilarity between the original and the despeckled images [93]:

$$Err = \left(\frac{1}{MN} \sum_{i=1}^{M} \sum_{j=1}^{N} |g_{i,j} - f_{i,j}|^{\beta} \right)^{1/\beta} \qquad (2.3)$$

computed for $\beta = 3$ (Err_3) and $\beta = 4$ (Err_4). For $\beta = 2$, the *RMSE* is computed as in (2.2), whereas for $\beta = 1$, the absolute difference, and for $\beta = \infty$ the maximum difference measure.

(d) The Geometric Average Error (*GAE*), is a measure, which shows if the despeckled image is very bad, it is used to replace or complete the *RMSE* and is computed as follows [98]:

$$GAE = \left(\prod_{i=1}^{M} \prod_{j=1}^{N} \sqrt{g_{i,j} - f_{i,j}} \right)^{1/MN}. \qquad (2.4)$$

The *GAE* is approaching zero, if there is a very good transformation (small differences) between the original and the despeckled image, and high vice versa. It is positive only if every pixel value is different between the original and the despeckled image. This measure is also used for tele-ultrasound, when transmitting ultrasound images. The *GAE* may be used to replace the *RMSE*, which is dominated by its large individual terms and is calculated for an image with dimensions $N \times M$. This amounts to a severe error in *RMSE* when large individual terms are present. For this reason the *RMSE* is often replaced by the *GAE*.

(e) While signal sensitivity and image noise properties are important by themselves, it is really the ratio of them that carries the most significance. The *SNR* is given by [99]:

$$SNR = 10\log_{10} \frac{\sum\limits_{i=1}^{M}\sum\limits_{j=1}^{N} \left(g_{i,j}^2 + f_{i,j}^2\right)}{\sum\limits_{i=1}^{M}\sum\limits_{j=1}^{N} \left(g_{i,j} - f_{i,j}\right)^2}. \tag{2.5}$$

It is calculated over an image area with dimensions $N \times M$. The *SNR*, *RMSE*, and *Err*, proved to be very sensitive tests for image degradation, but there are completely non-specific. Any small change, in image noise, despeckling, and transmitting preferences would cause an increase of the above measures.

(f) The peak *SNR* (*PSNR*) is computed using [99]

$$PSNR = -10\log_{10} \frac{MSE}{g_{\max}^2}, \tag{2.6}$$

where $g_{\max}^2$ is the maximum intensity in the unfiltered image. The *PSNR* is higher for a better-transformed image and lower for a poorly transformed image. It measures image fidelity that is how closely the despeckled image resembles the original image.

(g) The mathematically defined universal quality index [100] models any distortion as a combination of three different factors, which are: loss of correlation, luminance distortion, and contrast distortion and is derived as:

$$Q = \frac{\sigma_{gf}}{\sigma_f \sigma_g} \cdot \frac{2\bar{f}\bar{g}}{(\bar{f})^2 + (\bar{g})^2} \cdot \frac{2\sigma_f \sigma_g}{\sigma_f^2 + \sigma_g^2}, \qquad -1 < Q < 1, \tag{2.7}$$

where $\bar{g}$, and $\bar{f}$ represent the mean of the original and despeckled values with their standard deviations, σ_g, and σ_f, of the original and despeckled values of the analysis window, and σ_{gf} represents the covariance between the original and despeckled windows. Q is computed for a sliding window of size 8×8 without overlapping. Its highest value is 1 if $g_{i,j} = f_{i,j}$, while its lowest value is -1 if $f_{i,j} = 2\bar{g} - g_{i,j}$.

(h) The structural similarity index between two images [93], which is a generalization of (2.7), is given by:

$$SSIN = \frac{(2\bar{g}\bar{f} + c_1)(2\sigma_{gf} + c_2)}{(\bar{g}^2 + \bar{f}^2 + c_1)(\sigma_g^2 + \sigma_f^2 + c_2)}, \qquad -1 < SSIN < 1, \tag{2.8}$$

where $c_1 = 0.01dr$ and $c_2 = 0.03dr$, with $dr = 255$ representing the dynamic rage of the ultrasound images. The range of values for the *SSIN* lies between -1, for a bad and 1 for a good similarity between the original and despeckled images respectively. It is computed, similarly to the Q measure, for a sliding window of size 8×8 without overlapping.

(i) The speckle index, C, for log-compressed ultrasound images is defined as:

$$C = \frac{1}{MN} \sum_{i=1}^{M} \sum_{j=1}^{N} \frac{\sigma^2_{i,j}}{\mu_{i,j}} \tag{2.9}$$

and is an average measure of the amount of speckle presented in the image area with size $M \times N$, as a whole (over the whole image). It is used in most adaptive filters to adjust the weighting function $K_{i,j}$, in (3.2), described in Section 3.1, because it reflects the changes in contrast of the image in the presence of speckle noise. It does not depend on the intensity of the local mean but on the variance, σ^2, and the mean, μ, of the whole image. The larger C is, the more likely that the observed neighborhood belongs to an edge, thus C may be used also as an edge detector.

(j) Lesions detectability can be quantified using the contrast-to-speckle ratio, CSR [40, 42]. It is calculated by defining two regions of interest (i.e., the original image and the despeckled), and using the mean pixel value, and the variance, to quantify the contrast, $(\mu_g - \mu_f)/\mu_g$, and the speckle index noise, $\sqrt{(\sigma_g^2 + \sigma_f^2)}/\mu_g$. The ratio of these two quantities is termed as CSR and is defined as:

$$CSR = ((\bar{g} - \bar{f})\bar{g})/\sqrt{(\sigma_g^2 + \sigma_f^2)}, \tag{2.10}$$

where $\bar{g}$, $\bar{f}$, σ_g, σ_f, are the mean and standard deviations of the original and despeckle images respectively. The CSR, provides a quantitative measure of the detectability of low contrast lesions, when one region is completely inside the lesion and the second is the background media.

(k) The average difference (AD) between the original and the despeckled image defined as [101]:

$$AD = \sum_{i=1}^{M} \sum_{j=1}^{N} |g_{i,j} - f_{i,j}|. \tag{2.11}$$

(l) Structural content (SC) which is defined as [15]:

$$SC = \frac{\sum_{i=1}^{M} \sum_{j=1}^{N} g_{i,j}^2}{\sum_{i=1}^{M} \sum_{j=1}^{N} f_{i,j}^2}. \tag{2.12}$$

Large values of SC indicate images with poor quality.

(m) The normalized cross correlation (*NCC*) defined as [101]:

$$NCC = \frac{1}{MN} \frac{\sum\limits_{i=1}^{M} \sum\limits_{j=1}^{N} (g_{i,j} - \bar{g})(f_{i,j} - \bar{f})}{\sigma_g \sigma_f},$$
(2.13)

where M, N the image dimensions of g and f, $\bar{g}$, $\bar{f}$ the average of the original and de-speckled image, and σ_g, σ_f, the standard deviations of the original and despeckled images, respectively.

(n) The maximum difference (*MD*) as follows [101]:

$$MD = Max \left[\left| \sum\limits_{i=1}^{M} \sum\limits_{j=1}^{N} f_{i,j} - \sum\limits_{i=1}^{M} \sum\limits_{j=1}^{N} g_{i,j} \right| \right].$$
(2.14)

(o) The Laplacian mean square error (*LMSE*) which is based on the importance of edge measurement Large values of the *LMSE* indicate images with poor quality. The *LMSE* is defined as:

$$LMSE = \frac{\sum\limits_{i=1}^{M} \sum\limits_{j=1}^{N} \left[L\left(f_{i,j}\right) - L\left(g_{i,j}\right) \right]^2}{\sum\limits_{i=1}^{M} \sum\limits_{j=1}^{N} \left[L\left(f_{i,j}\right) \right]^2}$$
(2.15)

with $L(i, j)$ the Laplacian operator which is defined as [101]:

$$L(i, j) = g(i + 1, j) + g(i - 1, j) + g(i, j + 1) + g(i, j - 1) - 4g(i, j).$$

(p) The normalized absolute error (*NAE*) as [101]:

$$NAE = \frac{\sum\limits_{i=1}^{M} \sum\limits_{j=1}^{N} \left| g_{i,j} - f_{i,j} \right|}{\sum\limits_{i=1}^{M} \sum\limits_{j=1}^{N} \left| g_{i,j} \right|}.$$
(2.16)

Figure 2.2b illustrates a selected number of image quality metrics that are implemented both in IDF and VDF toolboxes.

The quality measures proposed above, do not necessarily correspond to all aspects of the expert's visual perception, nor do they correctly reflect structural coding artifacts [8], but if they are all combined together, and with the subjective tests, may offer a more accurate evaluation result. It is noted that subjective tests are tedious, time consuming, and expensive, and the results depend on the expert's background, motivation, and other factors [93, 98, 102, 103]. However, all these measures cover the visual quality just partially. The visual quality of image and video are difficult to define with mathematical precision, since they depend on the properties of our visual system. We know, for example, that our visual system is more tolerant to a certain amount of noise than to a reduced sharpness. On the other hand, it is very sensitive to certain specific artifacts, like blips and bumps [92].

CHAPTER 3

Linear Despeckle Filtering

This chapter provides the basic theoretical background of linear despeckle filtering techniques together with their algorithmic implementation MATLAB™ code for selected filters and practical examples on phantom and real ultrasound images. There are three groups of filters presented in this chapter, first order statistics filtering, local statistics filtering and homogeneous mask area filtering (see also Table 1.3). Despeckle filtering was evaluated for all filters presented in this book on phantom ultrasound carotid artery images (see Fig. 2.1) and real ultrasound images (see Fig. 1.7) and videos (see Fig. 7.6) of the CCA.

3.1 FIRST-ORDER STATISTICS FILTERING (*DSFLSMV, DSFWIENER*)

Most of the techniques for speckle reduction filtering in the literature use linear filtering based on local statistics. Their working principle may be described by a weighted average calculation using sub region statistics to estimate statistical measures over different pixel windows varying from 3×3 up to 15×15. All these techniques assume that the speckle noise model has a multiplicative form as given in (1.2) [3, 5, 7–13, 48, 78].

The filters utilizing the first-order statistics such as the variance and the mean of the neighborhood may be described with the model as in (1.4). Hence, the algorithms in this class may be traced back to the following equation [7, 13, 29, 50, 78] (see also Fig. 3.1):

$$f_{i,j} = \bar{g} + k_{i,j} \left(g_{i,j} - \bar{g} \right), \tag{3.1}$$

where $f_{i,j}$, is the estimated noise-free pixel value, $g_{i,j}$, is the noisy pixel value in the moving window, $\bar{g}$, is the local mean value of an $N_1 \times N_2$, region surrounding and including pixel $g_{i,j}$, $k_{i,j}$, is a weighting factor, with $k \in [0 \ldots 1]$, and i, j, are the pixel coordinates. The factor $k_{i,j}$, is a function of the local statistics in a moving window. It can be found in the literature [7, 13, 36, 38, 46, 78] and may be derived in different forms that:

$$k_{i,j} = \left(1 - \bar{g}^2 \sigma^2\right) / \left(\sigma^2 \left(1 + \sigma_n^2\right)\right) \tag{3.2}$$

$$k_{i,j} = \sigma^2 / \left(\bar{g}^2 \sigma_n^2 + \sigma^2\right) \tag{3.3}$$

$$k_{i,j} = \left(\sigma^2 - \sigma_n^2\right) / \sigma^2. \tag{3.4}$$

The values σ^2, and σ_n^2, represent the variance in the moving window and the variance of noise in the whole image respectively. The noise variance may be calculated for the logarithmically compressed image or video, by computing the average noise variance over a number of windows with dimensions considerable larger than the filtering window. In each window the noise variance is computed as:

$$\sigma_n^2 = \sum_{i=1}^{p} \sigma_p^2 / \bar{g}_p,$$

(3.5)

where σ_p^2 and $\bar{g}_p$ are the variance and mean of the noise in the selected windows, respectively, and p, is the index covering all windows in the whole image or video [3, 37, 40, 70]. If the value of $k_{i,j}$, is 1 (in edge areas) this will result to an unchanged pixel, whereas a value of 0 (in uniform areas) replaces the actual pixel by the local average, $\bar{g}$, over a small region of interest (see (3.1)). In this study the filter *DsFlsmv* uses Eq. (3.2).

The filter *DsFwiener* uses a pixel-wise adaptive wiener method [13, 32–35, 47, 78] implemented as given in (3.1), with the weighting factor $k_{i,j}$, as given in (3.4).

For both despeckle filters *DsFlsmv* and *DsFwiener* the moving window size was 5×5 (see also Fig. 3.1a).

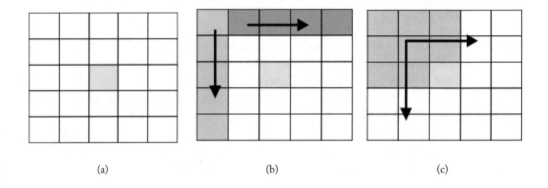

(a)   (b)   (c)

Figure 3.1: (a) Pixel moving window of 5×5 pixels, (b) schematical operation of the filters *DsFlsminv1d* with a 1D sliding moving window, and (c) *DsFlsmv* with a 2D sliding moving window.

Algorithm 3.1 presents the algorithmic steps for the implementation of the *DsFlsmv* despeckle filter, whereas Code 3.1 presents the implementation of the filter in MATLAB™ code for image despeckling. For video despeckling, the procedure described above is repeated for each consecutive video frame or selected video frames depending on the application.

Figure 3.2 illustrates the application of the despeckle filter *DsFlsmv* on the phantom image. Table 3.1 illustrates selected statistical and image quality metrics for the phantom image of Fig. 3.2 for the original and the despeckled images.

The despeckle filter *DsFlsmv* was applied on the phantom image varying the number of iterations (from 1–15) where the size of the sliding moving window was [5×5] as illustrated in Fig. 3.3. Table 3.2 tabulates the statistical and image quality features for the phantom image of

Algorithm 3.1 Linear Filtering: Linear Scaling Filter (*DsFlsmv*).

1	Load the image (first frame) for filtering
2	Specify the region of interest to be filtered, the moving window size (nhood) and the number of iterations (n)
3	Compute the noise variance σ_n^2 with (3.5) for the whole image (frame)
4	Starting from the left upper corner of the image (frame), compute for each moving window the coefficient $k_{i,j}$ in (3.2)
5	Compute $f_{i,j}$ in (3.1) and replace the noisy middle point in each moving window $g_{i,j}$, with the new computed value $f_{i,j}$
6	Repeat steps 4 and 5 for the whole image (frame) by sliding the moving window from left to right. For video despeckling repeat steps 1 to 6 for all consecutive video frames
7	Repeat steps 3 to 6 for n iterations
8	Compute the image quality evaluation metrics and the texture features for the original and the despeckled images (videos)
9	Display the original and despeckled images (or videos), the image (video) quality and evaluation metrics, and the texture features.

Table 3.1: Selected statistical and image quality features for Fig. 3.2 before and after despeckle filtering for the *DsFLsmv* filter. Bold values show improvement after despeckle filtering

Features	μ	Median	σ^2	σ^3	σ^4	Contrast	C	CSR	MSE	SNR	Q	SSIN	AD	SC	MD	NAE
Original	36	37	21	0.2	2.8	76	58	-	-	-	-	-	-	-	-	-
DsFlsmv	35	40	**17**	0.3	2.0	74	**48**	0.5	1757	28	0.5	0.54	0.7	1.1	74	0.22

C: Speckle Index, $C = (\sigma^2/\mu)100$, CSR: Contrast-speckle ratio, MSE: Mean square error, SNR: Peak signal-to-noise ratio, Q: Universal quality index, SSIN: Structrural similarity index, AD: Average difference, SC: Structureal content, MD: Maximum difference, NAE: Normalised absolute error.

Fig. 3.3, after the application of the *DsFlsmv* despeckle filter for increasing number of iterations. It is shown in Table 3.2, that increasing the number of iterations, the mean, median, skewness, σ^3, and kurtosis, σ^4, are preserved, whereas the standard deviation, σ^2, is slightly reduced. Furthermore, it is shown that increasing the numbers of iterations for the *DsFlsmv* filter, reduced contrast dramatically as demonstrated also in Fig. 3.3. Moreover, *C* is reduced with the number of iterations, whereas an increase of *C* for iterations 7, 8, and 9 was observed. The *CSR* increases slightly with the number of iterations. It is furthermore shown that the *SNR* is reduced significantly with increasing number of iterations. The *Q* and *SSIN* are reduced significantly with the

---

Code 3.1 Matlab™ Code Linear Filtering: Linear Scaling Filter (*DsFlsmv*). *(Continues.)*

```
        function f = DsFlsmv(g, nhood, niterations)
        % IDF toolbox, © Christos P. Loizou 2015
        %*********************************************************************
        % Local first order statistics filter
        % Input:
        % g:          Original (input) noisy image
        % nhood:      Size of the sliding moving window in pixels
        % Iterations: Number of iterations for which filtering is applied
        %
        % Output:
        % f:          Despeckled (output) image
        % Example:  f=DsFlsmv (g, [5 5], 5);
1       % Load the image for filtering
        %*********************************************************************
        imshow (g);                % show the original image
2       % Crop image region and select an area of interest to be despeckled
        % Specify the area of interest to be filtered, the moving window size (nhood) and the number of
        % iterations (n) the filtering is applied to the image
         [x, y, BW, xi, yi]=roipoly(g);
        maxx=max(xi); minx=min(xi); maxy=max(yi); miny=min(yi);
        [xsize, ysize, imagec, rect]=imcrop(g, [minx miny (maxx-minx) (maxy-miny)]);
        % the cropped image is  g = imagec
        % Check if the image loaded is a grayscale and normalize its values
        if isa(imagec, 'uint8')
          u8out = 1;
          if (islogical(imagec))
             logicalOut = 1;
           imagec = double(imagec);
        else
           logicalOut = 0;
           imagec = double(imagec)/255;
        end
        else
          u8out = 0;
        end
        % Calculate the noise and the standard deviation of the original image, and the noise variance in the
        % whole image
        % Compute the noise variance σ²ₙ with (3.5) from the whole image
3       stdnoise=(std2(imagec).*std2(imagec))/mean2(imagec);
        noisevar=stdnoise*stdnoise; %noise variance
        % Initialize a new image f (new image after filtering) with zeros
        f = imagec;
```

Code 3.1 *(Continued.)* Matlab™ Code Linear Filtering: Linear Scaling Filter (*DsFlsmv*). *(Continues.)*

| 4 | ```
% Apply n-iterations of the algorithm to the image
for i = 1:niterations
  fprintf('\rIteration %d',i);
  if i >=2
    imagec=f;
  end
% For each moving window, estimate the local mean of f.
localMean = filter2(ones(nhood), imagec) / prod(nhood);
% square of the local mean
lmsqr = localMean.*localMean;
% Starting from the left upper corner of the image, compute for each moving window the
% coefficient $k_{i,j}$ in (3.5)
localVar = filter2(ones(nhood), imagec.^2) / prod(nhood) - localMean.^2;
% Compute $f_{i,j}$ in (3.1) and replace the noisy middle point in each moving window $g_{i,j}$, with the
% new computed value $f_{i,j}$
f=localMean + (localVar - lmsqr .*noisevar ./ max(0.1, localVar + lmsqr .* noisevar)) .* (imagec - localMean);
``` |

```
end
% End for i Itterations
fprintf('\n');
if u8out==1,
  if (logicalOut)
    f = uint8(f);
  else
    f = uint8(round(f*255));
  end
end
```

6	% Repeat steps 4 and 5 for for the whole image by sliding the moving window from left to right
7	% Repeat steps 3 to 6 for n iterations specified
8	% Compute the texture and image quality evaluation metrics and display both the original and the

```
% despeckled images on the screen
% Calculate 61 Texture Features from the original and despeckled images
A=[ ]; F1=[ ] ;
%  Initialize the matrcies for texture features
T= DsTtexfeat(double(imagec));
A=[A, T'];
% Save the texture features of the original image in a matrix A
save or_texfeats A;

TAM=DsTtexfeat(double(f));
F1=[F1,TAM'];
save speckle1texfs F1;
```
```

Code 3.1 *(Continued.)* Matlab™ Code Linear Filtering: Linear Scaling Filter (*DsFlsmv*).

```
      % The texture features of the despeckled image are saved in matrix F1
      % Call the function metrics to calculate and display the 19 different image quality metrics between
      % the original and the despeckled image
      M=DsQmetrics(f, imagec);

9
      % Show original and despeckled images on the screen
      figure, imshow(imagec);
      figure, imshow(f);
```

Algorithm 3.1.1 presents the algorithmic steps for the implementation of the *DsFlsmv* despeckle filter, whereas Code 3.1.1 presents the implementation of the filter in MATLAB™ code for image despeckling.

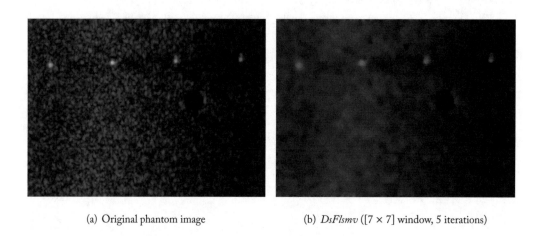

(a) Original phantom image   (b) *DsFlsmv* ([7 × 7] window, 5 iterations)

Figure 3.2: Example of *DsFlsmv* despeckle filtering on a phantom ultrasound image.

number of iterations. The despeckled phantom images of Fig. 3.3 were also visually assessed by the experts, where the best visual results were given for iterations 4, 5, and 6. For these iterations, the filtered image statistics remain the same, but contrast is further reduced.

The despeckle filter *DsFlsmv* was applied on the phantom image for different sliding moving window sizes (from [3 × 3] to [23 × 23]) where the number of iterations was kept constant to 5 as demonstrated in Fig. 3.4. Table 3.3 tabulates the statistical and image quality evaluation features for the phantom image after the application of the *DsFlsmv* filter illustrated in Fig. 3.4. It is shown that increasing the sliding moving window size, the mean and median are preserved and the variance is decreased for the window sizes [3 × 3] and [5 × 5].

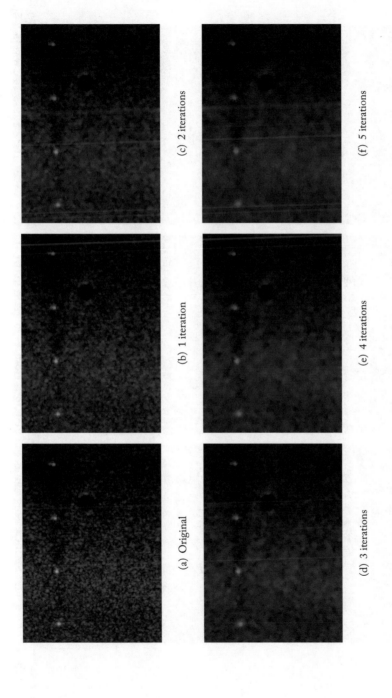

(a) Original

(b) 1 iteration

(c) 2 iterations

(d) 3 iterations

(e) 4 iterations

(f) 5 iterations

Figure 3.3: Original phantom ultrasound image given in (a) and the despeckled phantom images after the application of the *DsFlsmv* filter for increasing number of iterations from 1–15 given in (b)–(l), respectively. The window size was [5 × 5]. (*Continues.*)

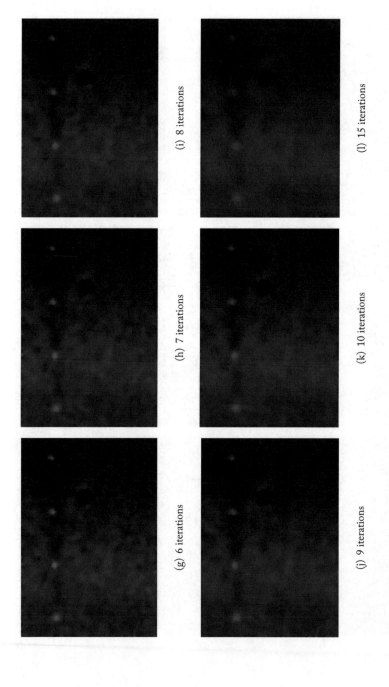

Figure 3.3: *(Continued.)* Original phantom ultrasound image given in (a) and the despeckled phantom images after the application of the *DsFlsmv* filter for increasing number of iterations from 1–15 given in (b)–(l), respectively. The window size was [5 × 5].

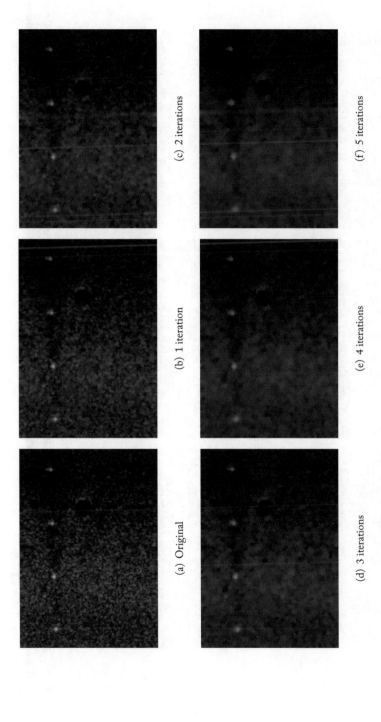

(a) Original

(b) 1 iteration

(c) 2 iterations

(d) 3 iterations

(e) 4 iterations

(f) 5 iterations

Figure 3.4:   Original phantom ultrasound image given in (a) and the despeckled phantom images after the application of the *DsFlsmv* filter for increasing pixel moving window size from [3 × 3] to [25 × 25] windows given in (b)–(l). The number of iterations was 5 for all cases. (*Continues.*)

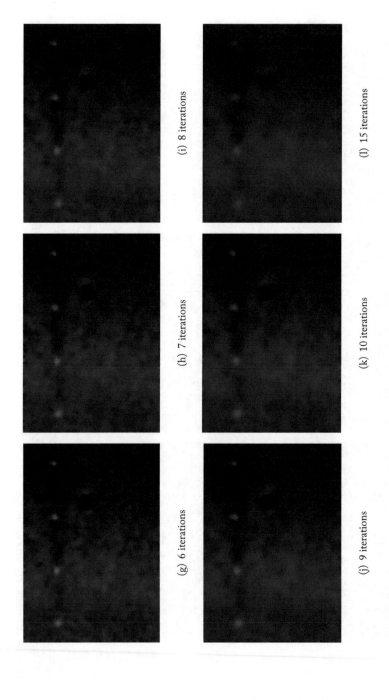

Figure 3.4: *(Continued.)* Original phantom ultrasound image given in (a) and the despeckled phantom images after the application of the *DsFlsmv* filter for increasing pixel moving window size from [3 × 3] to [25 × 25] windows given in (b)–(l). The number of iterations was 5 for all cases.

Table 3.2: Selected statistical and image quality evaluation features for Fig. 3.3 before and after despeckle filtering for the *DsFLsmv* filter for increasing the number of iterations and constant window size ([5 × 5])

FEATURES	NUMBER OF ITERATIONS												
	Original	1	2	3	4	5	6	7	8	9	10	15	20
μ	36	36	36	36	36	36	36	35	35	35	35	35	35
Median	37	38	39	39	40	40	40	40	40	40	40	40	39
σ^2	21	19	18	18	17	17	17	17	17	17	16	16	16
σ^3	0.2	0.01	0.2	0.2	0.3	0.3	0.3	0.3	0.3	0.4	0.4	0.4	0.4
σ^4	2.8	2.6	2.4	2.3	2.1	2.1	2	2	2	1.9	1.9	1.8	1.8
Contrast	76	35	17	9	6	3.8	2.9	2	1.7	1.4	1.3	0.8	0.6
C	58	53	50	50	47	47	47	49	49	49	46	46	46
CSR	-	0.43	0.43	0.44	0.44	0.44	0.44	0.45	0.45	0.45	0.46	0.46	0.46
MSE	-	1757	1758	1759	1760	1760	1762	1763	1764	1766	1778	1770	1771
SNR	-	21	17	18	16	15	14	13.7	13.3	12	12.5	12.9	12.5
Q	-	0.86	0.58	0.36	0.36	0.24	0.21	0.13	0.10	0.09	0.07	0.05	0.03
SSIN	-	0.88	0.71	0.55	0.55	0.51	0.49	0.47	0.45	0.43	0.42	0.41	0.39
MD	-	30	39	54	65	74	82	88	93	96	100	114	117
LMSE	-	0.58	0.88	0.91	0.95	1.02	0.99	0.99	0.99	1.01	1.01	1.01	1.02

Contrast is significantly reduced for window sizes [3 × 3] and [5 × 5], and then exhibits very small variations. Furthermore, it is shown that for increasing the number of iterations the filter *DsFlsmv* reduced the speckle index, *C*, especially for the first three window sizes and then it exhibits smaller variations, whereas the *CSR* remains constant. Table 3.3 also shows that the MSE remains relatively unchanged while the SNR is reduced significantly with increasing window size. The *Q* and *SSIN* are reduced significantly with the number of iterations. The despeckled phantom images of Fig. 3.4 were also visually assessed by the two experts, where the best visual results were given for the sliding window sizes of [3 × 3] and [5 × 5].

Figure 3.5 illustrates the application of the despeckle filter *DsFlsmv* on a video acquired from a symptomatic male subject at the age of 62 with 40% stenosis and a plaque at the far wall of the CCA. Despeckle filtering was applied on frames 1, 50, 100, 150, 200, and 250 for a moving window size of 5 × 5 pixels and 2 iterations.

The despeckle filter *DsFlsmv* was applied on the CCA ultrasound image of Fig. 3.5 for different number of iterations (from 1–20) where the sliding moving window was kept constant to [5 × 5] and the results are demonstrated in Table 3.4, where the statistical and image quality evaluation features for the application of the *DsFlsmv* filter are tabulated. It is shown that in-

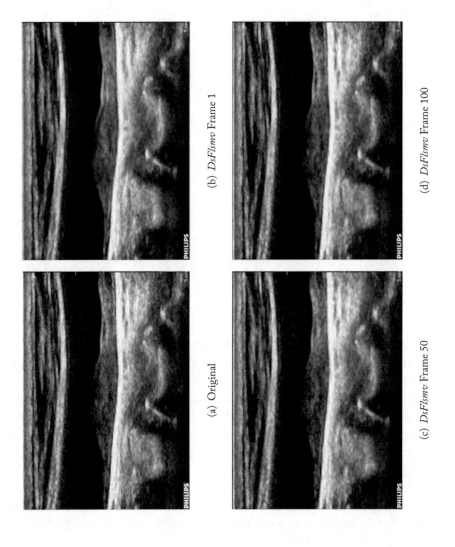

(a) Original

(b) *DsFlsmv* Frame 1

(c) *DsFlsmv* Frame 50

(d) *DsFlsmv* Frame 100

Figure 3.5: Examples of despeckle filtering with the filter *DsFlsmv* on a symptomatic video of the CCA with plaque at the far wall on frames 1, 50, 100, 150, 200, 250, and 300 illustrated in (b)–(h), respectively. *(Continues.)*

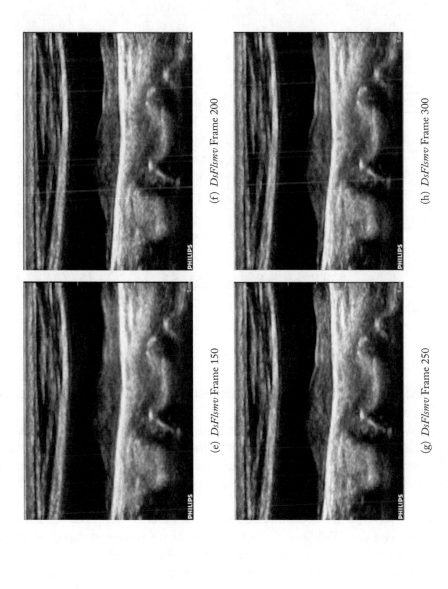

(e) *DsFlsmv* Frame 150

(f) *DsFlsmv* Frame 200

(g) *DsFlsmv* Frame 250

(h) *DsFlsmv* Frame 300

Figure 3.5: (*Continued.*) Examples of despeckle filtering with the filter *DsFlsmv* on a symptomatic video of the CCA with plaque at the far wall on frames 1, 50, 100, 150, 200, 250, and 300 illustrated in (b)–(h), respectively.

Table 3.3: Selected statistical and image quality evaluation features for Fig. 3.4 before and after despeckle filtering for the *DsFLsmv* filter for increasing the window size (from ([3 × 3]) to ([25 × 25]) window) and for five iterations

FEATURE	WINDOW SIZE												
	Original	x3	x5	x7	x9	x11	x13	x15	x17	x19	x21	x23	x25
μ	36	35	36	36	36	36	36	36	36	36	36	36	35
Median	37	38	39	40	40	41	41	41	41	41	41	41	41
σ^2	21	18	18	17	17	17	17	17	17	16	16	16	16
σ^3	0.2	0.06	0.2	0.3	0.3	0.4	0.4	0.4	0.4	0.4	0.4	0.4	0.5
σ^4	2.8	2.4	2.1	2	2	2	2	2	2	1.9	1.9	1.9	1.9
Contrast	76	14	6	4	3	3	3	4	4	4	5	5	6
C	58	51	50	47	47	47	47	47	47	44	44	44	46
CSR	-	0.44	0.44	0.44	0.44	0.44	0.44	0.44	0.44	0.44	0.44	0.44	0.46
MSE	-	1754	1757	1758	1760	1762	1764	1765	1766	1767	1768	1768	1770
SNR	-	19	16	15	14	14	13	12	12	11	11	10	9
Q	-	0.82	0.67	0.47	0.39	0.35	0.29	0.24	0.20	0.18	0.16	0.13	0.11
SSIN	-	0.80	0.61	0.59	0.57	0.56	0.54	0.52	0.50	0.49	0.48	0.46	0.45
MD	-	38	62	67	80	95	124	157	189	214	223	254	267
LMSE	-	0.78	0.93	0.94	0.99	2.5	3.4	4.5	5.6	6.2	7.1	8.02	8.3

creasing the number of iterations, the mean, median, and variance are preserved up to the third iteration and then are reduced.

Contrast and *SNR* are reduced with increasing the number of iterations. Furthermore, it is shown that for increasing the number of iterations the filter *DsFlsmv* increases the speckle index, *C*, and decreases the *CSR*. The *Q* and *SSIN* are reduced with the number of iterations. The despeckled phantom images of Fig. 3.4 were also visually assessed by the expert, where the best visual results were given for iterations 2 and 3.

Algorithm 3.2 presents the algorithmic steps for the implementation of the *DsFwiener* despeckle filter for image despeckling. For video despeckling the procedure described above is repeated for each consecutive video frame or selected video frames depending on the application.

Figure 3.6 illustrates the application of the despeckle filter *DsFwiener* on the phantom image of the CCA with the original and the despeckled images shown on the left and right column, respectively. Table 3.5 illustrates selected statistical and image quality features for the phantom image of Fig. 3.6 for the original and despeckled images.

Table 3.4: Selected statistical and image quality evaluation features for Fig. 3.5 before and after despeckle filtering with the *Ds-FLsmv* filter for increasing the number of iterations and constant window size of ([5 × 5])

FEATURES	Original	NUMBER OF ITERATIONS											
		1	2	3	4	5	6	7	8	9	10	15	20
μ	77	76	76	75	75	73	72	71	70	69	68	65	45
Median	68	68	68	67	65	64	63	62	62	60	59	57	49
σ^2	67	64	63	62	61	60	58	57	55	54	53	46	39
σ^3	0.63	0.58	0.56	0.53	0.49	0.47	0.44	0.41	0.39	0.37	0.36	0.31	0.26
σ^4	2.4	2.4	2.4	2.4	2.5	2.5	2.6	2.6	2.7	2.7	3.0	3.2	3.6
Contrast	334	313	310	308	307	305	303	300	296	292	286	272	243
C	87	89	89	90	91	93	35	97	98	100	102	115	126
CSR	-	0.09	0.09	0.08	0.08	0.07	0.07	0.07	0.06	0.06	0.05	0.03	0.03
MSE	-	10480	10482	10485	10490	10493	10496	10499	10501	10555	10555	10560	10656
SNR	-	25	25	24	22	21	20	19	18	17	15	12	10
Q	-	0.87	0.85	0.81	0.78	0.75	0.72	0.68	0.63	0.54	0.46	0.39	0.27
SSIN	-	0.89	0.86	0.83	0.79	0.77	0.73	0.70	0.62	0.52	0.47	0.38	0.26
MD	-	167	171	182	189	192	198	203	212	225	230	245	267
LMSE	-	0.18	0.32	0.35	0.41	0.47	0.54	0.67	0.74	0.88	0.92	1.92	2.01

Algorithm 3.2 Linear Filtering: Linear Scaling Filter (*DsFwiener*).

---

1   Load the image for filtering

2   Specify the region of interest to be filtered, the moving sliding window size and the number of iterations

3   Compute the noise variance σ_n^2 with (3.5) for the whole image (frame)

4   Starting from the left upper corner of the image (frame), compute for each sliding moving window the coefficient $k_{i,j}$ in (3.4)

5   Compute $f_{i,j}$ in (3.1) and replace the noisy middle point in each moving window $g_{i,j}$, with the new computed value $f_{i,j}$

6   Repeat steps 4 and 5 for all the pixels in the image (frame) by sliding the moving window from left to right

For video despeckling repeat steps 1 to 6 for all consecutive video frames.

7   Repeat steps 3 to 6 for a second iteration of despeckle filtering

8   Compute the image quality evaluation metrics and the texture features for the original and the despeckled images (video frames)

9   Display the original and despeckled images (videos), the image quality and evaluation metrics, and the texture features.

Table 3.5: Selected statistical and image quality features for Fig. 3.6 before and after despeckle filtering for the *DsFwiener* despeckle filter. Bolded values show improvements after despeckle filtering

Features	μ	Median	σ^2	σ^3	σ^4	Contrast	C	CSR	MSE	PSNR	Q	SSIN	AD	SC	MD	NAE
Original	36	37	21	0.2	2.8	76	58	-	-	-	-	-	-	-	-	-
DsFwiener	**36**	39	**18**	0.06	3.0	65	**50**	10	1730	29	0.4	0.58	0.1	1.1	34	0.21

3.2   LOCAL STATISTICS FILTERING WITH HIGHER MOMENTS (*DSFLSMINV1D, DSFLSMVSK2D*)

As discussed earlier many of the despeckle filters proposed in the literature suffer from smoothing effects in edge areas. Because of their statistical working principle, the edges may be better detected by incorporating higher statistical variance moments (variance, skewness, kurtosis) [61], calculated from the local moving window. The variance in every window, σ_w^2, may thus be described as a function of the variance, σ^2, skewness, σ^3, and kurtosis, σ^4, in the sliding moving

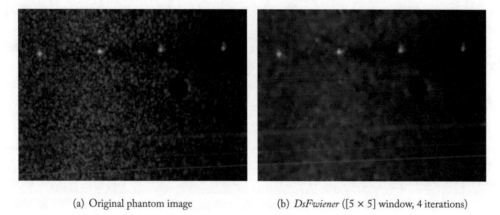

(a) Original phantom image   (b) *DsFwiener* ([5 × 5] window, 4 iterations)

Figure 3.6: Example of *DsFwiener* despeckle filtering on a phantom ultrasound image.

local window, and is calculated for the filter *DsFlsminv1d* as (see also Fig. 3.1b):

$$\sigma_w^2 = \left(c_2\sigma^2 + c_3\sigma^3 + c_4\sigma^4\right) / \left(c_2 + c_3 + c_4\right). \tag{3.6}$$

The constants c_2, c_3, c_4, in (3.6) may be calculated using [7]:

$$R = 1 - \frac{1}{1 + \sigma^2} \tag{3.7}$$

which represents the smoothness of the image. Specifically, the constants, c_2, c_3, c_4, are calculated, by replacing the variance σ^2, in (3.7), the skewness, σ^3, and the kurtosis, σ^4, in the moving pixel window, respectively. The higher moments are each, weighted with a factor, c_2, c_3, c_4, which receives values, $0 < c < 1$. Equations (3.6) and (3.7), will be applied in windows where.

$$c_3\sigma^3 \le c_2\sigma^2 \le c_4\sigma^4. \tag{3.8}$$

In regions where (3.8) is not valid, the window variance can be calculated as:

$$\sigma_w^2 = \left(c_2\sigma^2 + c_4\sigma^4\right) / \left(c_2 + c_4\right). \tag{3.9}$$

The final value for the σ_w^2, will be used to replace the variance, σ^2, and will be further used for calculating the coefficient of variation in (3.4). The *DsFlsminv1d* despeckle filter operates in the 1D direction, by calculating the σ_w^2 for each row and each column in the sliding moving window (see Fig. 2.1b), where the introduction of the higher moments in the filtering process should preserve the edges and should not smooth the image in areas with strong pixel variations. The middle pixel in the window is then replaced with (3.6), by replacing the $k_{i,j}$ weighting factor with the σ_w^2. The σ_w^2 in (3.9), can be interpreted as a generalized moment weighting factor with the weighting coefficients c_2, c_3, c_4. The moving window size for the *DsFlsminv1d* filter was 5×5 and its operation is shown in Fig. 2.1b.

The despeckle filter *DsFlsmvsk2d* [7], is the 2D realization of the *DsFlsminv1d* utilizing the higher statistical moments, σ^3 and σ^4, of the image in a 5×5 pixel moving window.

Algorithm 3.3 presents the algorithmic steps for the implementation of the *DsFlsmvsk2d* despeckle filter.

Algorithm 3.3 Linear Filtering: Linear Scaling Filter (*DsFlsmvsk2d*).

1	Load the image for filtering
2	Specify the region of interest to be filtered, the moving window size and the number of iterations (n)
3	Compute the noise variance σ_n^2 with (3.5) for the whole image
4	Starting from the left upper corner of the image, compute for each moving window the coefficient *win_* var as follows:
5	If (3.8) is true, use (3.8) otherwise use (3.9)
6	Compute $f_{i,j}$ in (3.1) and replace the noisy middle point in each moving window $g_{i,j}$, with the new computed value $f_{i,j}$, by using the *win_* var for the coefficient of variation $k_{i,j}$
7	Repeat steps 4 to 6 for all the pixels in the image by sliding the moving window from left to right
8	Repeat steps 3 to 7 for n iterations
9	Compute the image quality evaluation metrics and the texture features for the original and the despeckled images
10	Display the original and despeckled images, the image quality and evaluation metrics, and the texture features.

Figure 3.7 illustrates the application of the despeckle filters *DsFlsminv1d* and *DsFlsmvsk2d* on the phantom ultrasound image of the CCA with the original shown in Fig. 3.7a and the despeckled images shown in Fig. 3.7b and Fig. 3.7c, d, respectively. Table 3.6 illustrates selected statistical and image quality features for the phantom image of Fig. 3.7 for the original and despeckled images. Bolded values show improvement after despeckle filtering. It is shown that both filters preserve the mean, while the *DsFlsmvsk2d* increases enormously the *CSR*.

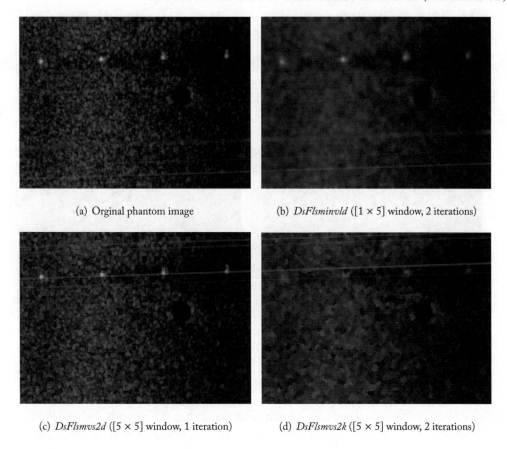

(a) Orginal phantom image

(b) *DsFlsminv1d* ([1 × 5] window, 2 iterations)

(c) *DsFlsmvs2d* ([5 × 5] window, 1 iteration)

(d) *DsFlsmvs2k* ([5 × 5] window, 2 iterations)

Figure 3.7: Example of *DsFlsminv1d* in (b), and *DsFLsmvsk2d* despeckle filtering on a phantom ultrasound image of the CCA in (c) and (d), respectively.

3.3 HOMOGENEOUS MASK AREA FILTERING (*DSFLSMINSC*)

The *DsFlsminsc* is a 2D filter operating in a 5 × 5 pixel neighborhood by searching for the most homogenous neighborhood area around each pixel, using a 3 × 3 subset window [79], as shown in Fig. 3.1c. The middle pixel of the 5 × 5 neighborhood is substituted with the average gray level of the 3 × 3 mask with the smallest speckle index, C, where C for log-compressed images is given in (2.9). The window with the smallest C is the most homogenous semi-window, which presumably, does not contain any edge. The filter is applied iteratively until the gray levels of almost all pixels in the image do not change.

The operation of the *DsFlsminsc* filter may be described as follows (see also Fig. 3.1c):

Table 3.6: Selected statistical features for Fig. 3.7 before and after despeckle filtering for the *Ds-FLsminv1d* and *DsFLsmvsk2d* (for 1 iteration) despeckle filters

Features	μ	Median	σ^2	σ^3	σ^4	Contrast	C	CSR	MSE	PSNR	Q	SSIN	AD	SC	MD	NAE
Original	36	37	21	0.2	2.8	76	58	-	-	-	-	-	-	-	-	-
DsFlsminv1d	**34**	36	19	0.12	2.0	54	**56**	3	1746	16	0.5	0.65	0.09	1.2	44	0.21
DsFlsmvsk2d	**36**	43	18	0.19	2.1	49	**52**	10	1735	14	0.5	0.79	1.2	1.1	1112	0.20

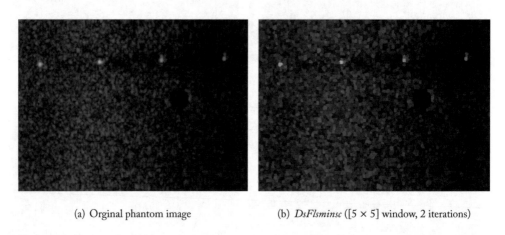

(a) Orginal phantom image   (b) *DsFlsminsc* ([5 × 5] window, 2 iterations)

Figure 3.8: Example of the *DsFlsminsc* despeckled filtering on a phantom ultrasound image.

Table 3.7: Selected statistical and image quality features for Fig. 3.8 before and after despeckle filtering for the *DsFLsminsc* despeckle filter. Bold values show improvement after despeckle filtering

Features	μ	Median	σ^2	σ^3	σ^4	Contrast	C	CSR	MSE	SNR	Q	SSIN	AD	SC	MD	NAE
Original	36	37	21	0.2	2.8	76	58	-	-	-	-	-	-	-	-	-
DsFlsminsc	36	36	22	0.2	2.8	75	**48**	21	1757	20	0.84	0.86	0.22	0.98	54	0.13

(a) slide the 3 × 3 mask with the (5 × 5 pixel) selected window;

(b) detect the position of the mask for which the *C* (3.3) is minimum;

(c) assign the average gray level of the mask to the middle pixel of the 5 × 5 window;

(d) apply steps (a) to (c) for all pixels in the image; and

(e) iterate the above process until the gray levels of almost all pixels in the image do not change.

Algorithm 3.4 presents the algorithmic steps for the implementation of the *DsFlsminsc* despeckle filter, whereas Code 3.2 presents the implementation of the filter in MATLAB™ code.

Figure 3.8 shows the application of the despeckle filter *DsFlsminsc* on a phantom ultrasound image for a moving window size 5 × 5 pixels and 3 iterations.

Table 3.7 illustrates selected statistical and image quality features for the phantom image of Fig. 3.8, the original and despeckled images. Bolded values show improvement after despeckle filtering. It is shown that filter *DsFlsminsc* preserve almost all features very well and decreases C.

Algorithm 3.4 Linear Filtering: Homogeneous Mask Area Filtering (*DsFlsminsc*).

1	Load the image for filtering
2	Specify the region of interest to be filtered, the moving window size, the number of iterations (n) and the edge detector to be used
3	Starting from the left upper corner of the image, rotate a mask around the middle pixel of the window for each moving window
4	Detect the position of the mask for which C (2.9) is minimum
5	Assign the average gray level of the mask at the selected position to the middle pixel in the 5x5 window
6	Repeat steps 4 and 5 for all the pixels in the image by sliding the moving window from left to right
7	Repeat steps 3 to 6 for a second iteration of despeckle filtering
8	Compute the image quality evaluation metrics and the texture features for the original and the despeckled images
9	Display the original and despeckled images, the image quality and evaluation metrics, and the texture features.

Code 3.2 Matlab™ Code Linear Filtering: Homogeneous Mask Area Filtering (*DsFlsminsc*). *(Continues.)*

```
1   function f = DsFlsminsc(g, nhood, niterations, edge)
    %**********************************************************************
    % Despeckle filtering toolbox, % © Christos P. Loizou 2015
    % Ultrasound image-Multiplicative noise filtering
    % The filter utilizes different filter detectors, from which you may choose one according to your
    % application
    % Input variables:
    % g :          input image to be filtered, i.e. 'cell.tif'
    % nhood :      sliding moving window, i.e [5 5]
    % niterations:   iterations for which filtering is applied iteratively
    % edge   :edge detector, used for finding the most homogeneous areas within the sliding window
    %          : edge=0, use the variance as an edge detector
    %          : edge=1, use the speckle contrast as an edge detector
    %          : edge=2, use max|m1-m2| input 2, max|m1/m2, m2/m1|'as an edge detector
    %          : edge=3, use the third moment as an edge detector
    %          : edge=4, use the fourth moment as an edge detector
    %
    % Output variable:
    % f        :input image for filtering
    %**********************************************************************
    % Specify the area of interest to be filtered, the moving window size, the number of times
2   % (niterations) the filtering is applied to the image and the edge detector to be used
    disp('Input the edge detector you would like to be used for the filter..');
    disp('Input 0 for using the variance as a detector');
    disp('Speckle Contrast input 1, max|m1-m2| input 2, max|m1/m2, m2/m1|');
    disp('Moment 3rd grades 3, Moment 4th grades 4');
    if isa(g, 'uint8')
      u8out = 1;
      if (islogical(g))
        logicalOut = 1;
        g = double(g);
      else
        logicalOut = 0;
        g = double(g)/255;
      end
    else
      u8out = 0;
    end
    % Estimate the size of the image
    [ma ,na] = size(g);
    % Estimate the midle of the processing window, which takes onle values 3, 5, 7
    z=(nhood(1)-1)/2;
```

Code 3.2 *(Continued.)* Matlab™ Code Linear Filtering: Homogeneous Mask Area Filtering (*Ds-Flsminsc*). *(Continues.)*

```
% Initialize the picture f (new picture) with zeros
f=g;
% Apply the filter niterations on the original image
  for i = 1:niterations
  fprintf('\rIteration %d',i);
  if i >=2
    g=f;
  end
% Starting from the left upper corner of the image, rotate a mask around the middle pixel of the
% window for each moving window
% Estimate and change the middle pixel in the sliding window
handle=waitbar(0, 'Calculating/replacing the center pixel in a sliding window...');
ini=z+1;
for i= ini :(ma-z)
  for j= ini:(na-z)
    var_neu=1000000000.0;si_neu=10000000000.0; xmit1=0.0; cd_neu=10000000.0;
    hos_neu = 1000000000.0; hos4_neu = 100000000.0;
    for a= (i-z):i
      for b=(j-z):j
        xmit= 0.0;
        for l=a:(a+z)
          for p=b:(b+z)
            xmit=xmit + g(l, p);
          end
% End for p
        end
% End for l
        xmit = (1.0/9.0) *xmit;
        var=0.0; pk=0.0; pk4=0.0;
% Detect the position of the mask for which the C of the gray levels is minimum
        for l=a:(a+z)
          for p=b:(b+z)
            var= var + ((g(l,p)-xmit)*(g(l, p)-xmit) );
            % 3rd moment
            pk= pk + ((g(l,p)-xmit)*(g(l, p)-xmit) * (g(l, p)-xmit) );
            pk4=pk4 +(g(l,p)-xmit)*(g(l, p)-xmit)*(g(l, p)-xmit)*(g(l,p)-xmit);
          end
% End for p
        end
% End for l
% Variance in subwindow
        var = (1/9.0)* var;
```

3

4

Code 3.2 *(Continued.)* Matlab™ Code Linear Filtering: Homogeneous Mask Area Filtering (*Ds-Flsminsc*). *(Continues.)*

```
            % 3rd moment in window
                pk = (1/9.0)*pk;
            % Assign the average gray level of the mask at the selected position to the middle pixel
5               if xmit ~=0.0
            % Speckle index in subwindow
                si = sqrt(var)/xmit;
                else
                si=0.0;
                end
            % Gradient information of the subset
                cd = abs(xmit-xmit1);
                xmit1 = xmit;
                if xmit~=0.0
            % 3rd higher order statistics
                hos = power(pk, 0.5) /xmit;
                % 4th higher order statistics
                hos4 = power (pk4, 0.25)/xmit;
                else
                hos=0.0;
                hos4=0.0;
                end
                % Use the speckle contrast to calculate f(i, j)
                if edge == 1
                  if si < si_neu
                  si_neu = si;
                  f(i, j) = xmit;
                  end
                % end if speckle index
                elseif edge == 0
            % Use the variance to calculate f(i, j)
                if var <var_neu
                var_neu = var;
                f(i, j) = xmit;
                end
                % end if var
                elseif edge == 2
                if cd < cd_neu
                  cd_neu =cd;
            % Use the local gradient to calculate f(i, j)
                  f(i, j)= xmit;
                end
```

Code 3.2 *(Continued.)* Matlab™ Code Linear Filtering: Homogeneous Mask Area Filtering (*Ds-Flsminsc*). *(Continues.)*

```
     % End if local gradient
           elseif edge == 3
             if hos < hos_neu
               hos_neu = hos;
     % Use higher moments to calculate f(i, j)
               f(i, j) = xmit;
             end
     % end if higher order statistics 3rd grades
           elseif edge == 4
             if hos4 < hos4_neu
               hos4_neu = hos4;
     % use higher moments to 4th grades to calculate f(i, j)
               f(i, j) = xmit;
             end                 % end if higher order statistics 4th grades
           end                   % end if edge

         end   % end for b
       end     % end for a
     % Repeat steps 4 and 5 for all the pixels in the image by sliding the moving window from left to
     % right
      end    %e nd for n
        waitbar(i/na)
     end     % end for m
6    % Repeat steps 3 to 6 for a second iteration of despeckle filtering
     close(handle)

     end
     % End for itterations
     fprintf('\n');

7    if u8out==1,
      if (logicalOut)
       f = uint8(f);
      else
       f = uint8(round(f*255));
      end
     end
     % Calculate 61 Texture Features from the original and despeckled images
8    A=[ ]; F1=[ ] ;
     % Initialize the matrcies for texture features
     T= DsTtexfeat(double(g));
```

Code 3.2 *(Continued.)* Matlab™ Code Linear Filtering: Homogeneous Mask Area Filtering (*Ds-Flsminsc*).

```
A=[A, T'];
% Save the texture features of the original image in a matrix A
save or_texfeats A;

TAM=DsTtexfeat(double(f));
F1=[F1,TAM'];
save speckle1texfs F1;
% The texture features of the despeckled image are saved in matrix F1
% Call the function metrics to calculate and display the 19 different image quality metrics between
%  the original and the despeckled image
M=DsQmetrics(f, g);

% Display both the original and the despeckled images on the screen
figure, imshow(g), title('Original Image');
figure, imshow(f), title('Image filtered by maskedge filter');
```

CHAPTER 4

Nonlinear Despeckle Filtering

This chapter provides the basic theoretical background of nonlinear despeckle filtering techniques together with their algorithmic implementation MATLAB™ code for selected filters and practical examples on phantom ultrasound images. There are eight different filters presented in this chapter which are the following: median filtering (*DsFmedian*), linear scaling filtering (*DsFca*, *DsFlecasort, DsFls*), maximum homogeneity over a pixel neighborhood (*DsFhomog*), geometric filtering (*DsFgf4d*), homomorphic filtering (*DsFhomo*), hybrid median filtering (*DsFhmedian*), Kuwahara filtering (*DsFkuwahara*), and non-local filtering (*DsFnlocal*) (see also Table 1.3). Nonlinear filtering is based on nonlinear operations involving the pixels a neighborhood. For example, letting the center pixel in the moving window be equal to the maximum pixel in its neighborhood is a nonlinear filtering operation.

4.1 MEDIAN FILTERING (*DSFMEDIAN*)

The filter *DsFmedian* [60] is a simple nonlinear operator that replaces the middle pixel in the window with the median-value of its neighbors. Algorithm 4.1 presents the algorithmic steps for the implementation of the *DsFmedian* despeckle filter.

Figure 4.1 shows the application of the despeckle filter *DsFmedian* on the phantom image for a moving window size 5 × 5 pixels and 4 iterations. Table 4.1 illustrates selected statistical and image quality features for the phantom image of Fig. 4.1 for the original and the despeckled images. It is shown that the *DsFmedian* filter preserves the mean, decreases slightly the variance, while decreases the C.

Table 4.1: Selected statistical and image quality features for Fig. 4.1 before and after despeckle filtering for the *DsFmedian* despeckle filter. Bold values show improvement after despeckle filtering

Features	μ	Median	σ^2	σ^3	σ^4	Contrast	C	CSR	MSE	PSNR	Q	SSIN	AD	SC	MD	NAE
Original	36	37	21	0.2	2.8	76	58	-	-	-	-	-	-	-	-	-
DsFmedian	36	39	**19**	1.2	14.3	67	**53**	2	1740	27	0.37	0.57	0.53	1.2	134	0.23

Algorithm 4.1 Nonlinear Filtering: Median Filter (*DsFmedian*).

---

1   Load the image for filtering

2   Specify the region of interest to be filtered, the moving window size and the number of iterations (n)

3   Starting from the left upper corner of the image, compute for each sliding moving window its median value

4   Replace the middle pixel in the sliding window with the median value calculated in step 3

5   Repeat steps 3 and 4 for the whole image by sliding the moving window from left to right

6   Repeat steps 3 to 5 for n iterations

7   Compute the image quality evaluation metrics and the texture features for the original and the despeckled images

8   Display the original and despeckled images, the image quality and evaluation metrics, and the texture features.

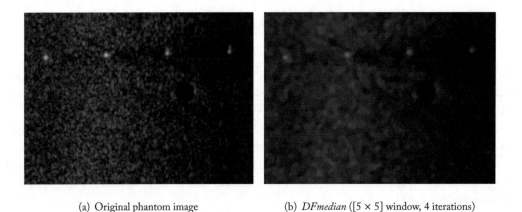

(a) Original phantom image   (b) *DFmedian* ([5 × 5] window, 4 iterations)

Figure 4.1: Example of the *DsFmedian* despeckle filtering on a phantom ultrasound image.

4.2   LINEAR SCALING FILTER *(DSFCA, DSFLECASORT, DSFLS)*

The *DsFca* filter despeckles the image through linear scaling of the gray level values [19]. In a window of 5 × 5 pixels, compute the mean of all pixels whose difference in the gray level with the intensity $g_{i,j}$, (middle pixel in the moving window), is lower or equal to a given threshold ϑ. Assign this value to the gray level $g_{i,j}$ with $\vartheta = \alpha * g_{max}$, where g_{max}, is the maximum gray level of the image and $\alpha = [0, 1]$. Best results can be obtained with $\alpha = 0.1$.

The *DsFlecasort* filter [19] takes k points of a pixel neighborhood, which are closest to the gray level of the image at point $g_{i,j}$, (middle point in the moving window) including $g_{i,j}$ [69]. It then assigns the mean value of these points to the pixel $g_{i,j}$. Usually, $N = 9$ in a 3×3 window, where $k = 6$.

The *DsFls* filter [47], scales the pixel intensities by finding the maximum, g_{max}, and the minimum, g_{min}, gray level values in every moving window and then replaces the middle pixel with:

$$f_{i,j} = \frac{g_{max} + g_{min}}{2}. \tag{4.1}$$

Algorithm 4.2 presents the algorithmic steps for the implementation of the *DsFca* despeckle filter.

Algorithm 4.2 Nonlinear Filtering: Linear Scaling Filtering (*DsFca*) Filter.

1	Load the image for filtering
2	Specify the region of interest to be filtered, the moving window size and the number of iterations (n)
3	Starting from the left upper corner of the image, compute for each moving window the mean of all pixels in the window whose difference in the gray level with the middle pixel in the moving window, is lower or equal to a given threshold ϑ, with $\vartheta = \alpha * g_{max}$, where g_{max}, is the
4	Assign the computed mean value to the middle pixel in the window
5	Repeat steps 3 and 4 for n iterations
6	Compute the image quality evaluation metrics and the texture features for the original and the despeckled images
7	Display the original and despeckled images, the image quality and evaluation metrics, and the texture features.

Figure 4.2 illustrates the application of the despeckle filters *DsFca*, *DsFlecasort*, and *DsFls* on a phantom ultrasound image. Table 4.2 presents selected texture features and image quality evaluation metrics for the phantom image of Fig. 4.2 for the original and the despeckled images. It is shown that all three despeckle filters preserve the mean, reduce slightly the variance, reduce significantly the contrast and reduce C.

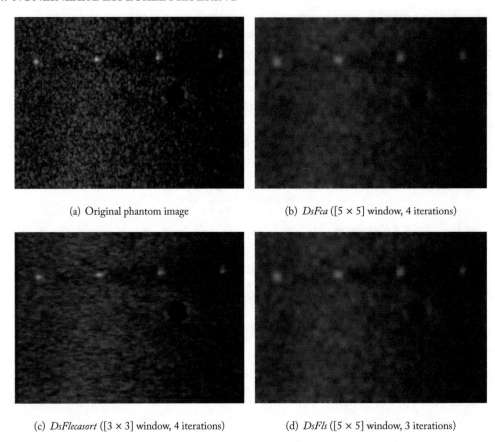

(a) Original phantom image   (b) *DsFca* ([5 × 5] window, 4 iterations)

(c) *DsFlecasort* ([3 × 3] window, 4 iterations)   (d) *DsFls* ([5 × 5] window, 3 iterations)

Figure 4.2: Examples of (b) *DsFca*, (c) *DsFlecasort*, and (d) *DsFls* despeckle filtering on a phantom image shown in (a).

4.3 MAXIMUM HOMOGENEITY OVER A PIXEL NEIGHBORHOOD FILTERING (*DSFHOMOG*)

The *DsFhomog* filter is based on an estimation of the most homogeneous neighborhood around each image pixel [54]. The filter takes into consideration only pixels that belong in the processed neighborhood (7 × 7 pixels) using (4.2), under the assumption that the observed area is homogeneous. The output image is then given by:

$$f_{i,j} = (c_{i,j} g_{i,j}) / \sum_{i,j} c_{i,j} \quad \text{with} \quad \begin{cases} c_{i,j} = 1 & \text{if} \quad (1 - 2\sigma_n)\bar{g} \leq g_{i,j} \leq (1 + 2\sigma_n)\bar{g} \\ c_{i,j} = 0 & \text{otherwise.} \end{cases} \quad (4.2)$$

Table 4.2: Selected statistical and image quality features for Fig. 4.2 before and after despeckle filtering for the *DsFca*, *DsFlecasort*, and *DsFls* despeckle filters. Bold values show improvement after despeckle filtering

Features	μ	Median	σ^2	σ^3	σ^4	Contrast	C	CSR	MSE	PSNR	Q	SSIN	AD	SC	MD	NAE
Original	36	37	21	0.2	2.8	76	58	-	-	-	-	-	-	-	-	-
DsFca	37	39	**19**	0.2	2.3	6	**51**	15	1760	27	0.21	0.41	-0.11	1.10	73	0.27
DsFlecasort	36	38	**18**	0.03	2.4	22	**50**	2	1754	22	0.22	0.41	0.52	1.05	94	0.30
DsFls	37	40	**17**	0.16	2.3	5	**46**	5	1765	25	0.27	0.51	-0.58	1.08	69	0.24

The *DsFhomog* filter does not require any parameters or thresholds to be tuned, thus making the filter suitable for automatic implementation.

Algorithm 4.3 presents the algorithmic steps for the implementation of the *DsFhomog* despeckle filter.

Algorithm 4.3 Nonlinear Filtering: Maximum Homogeneity Filter (*DsFhomog*).

---

1   Load the image for filtering

2   Specify the region of interest to be filtered, the moving window size and the number of iterations (n)

3   Starting from the left upper corner of the image, apply for each moving window equation (4.2) and replace the middle pixel with the new value

4   Repeat step 3 for the whole image by sliding the moving window from left to right

5   Repeat steps 3 to 4 for n iterations

6   Compute the image quality evaluation metrics and the texture features for the original and the despeckled images

7   Display the original and despeckled images, the image quality and evaluation metrics, and the texture features.

Table 4.3: Selected statistical and image quality features for Fig. 4.3 before and after despeckle filtering for the *DsFhomog* despeckle filter. Bold values show improvement after despeckle filtering

Features	μ	Median	σ^2	σ^3	σ^4	Contrast	C	CSR	MSE	PSNR	Q	SSIN	AD	SC	MD	NAE
Original	36	37	21	0.2	2.8	76	58	-	-	-	-	-	-	-	-	-
DsFhomog	36	39	**17**	0.04	3.1	9	**47**	45	1760	30	0.51	0.67	0.20	1.09	146	0.19

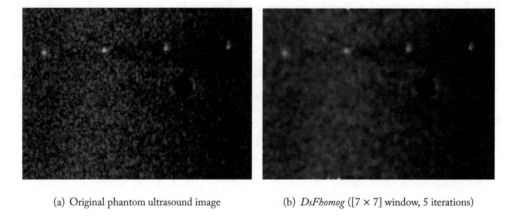

(a) Original phantom ultrasound image   (b) *DsFhomog* ([7 × 7] window, 5 iterations)

Figure 4.3: Example of the *DsFhomog* despeckle filtering on a phantom ultrasound image.

Figure 4.3 shows the application of the despeckle filter *DsFhomog* on a phantom ultrasound image for a moving window size 7 × 7 pixels and 5 iterations. Table 4.3 illustrates selected statistical and image quality features for the phantom image of Fig. 4.3 for the original and the despeckled images. It is shown that the *DsFhomog* filter preserves the mean, reduces the variance and decreases the C and decreases significantly the contrast of the image.

4.4   GEOMETRIC FILTERING (*DSFGF4D*)

The concept of the geometric filtering is that speckle appears in the image as narrow walls and valleys. The geometric filter, through iterative repetition, gradually tears down the narrow walls (bright edges) and fills up the narrow valleys (dark edges), thus smearing the weak edges that need to be preserved.

The *DsFgf4d* filter [44] uses a nonlinear noise reduction technique. It compares the intensity of the central pixel in a 3 × 3 neighborhood with those of its 8 neighbors, and based upon the neighborhood pixel intensities it increments or decrements the intensity of the central pixel such that it becomes more representative of its surroundings. The operation of the geometric filter *DsFgf4d* may be described with Fig. 4.4a and Fig. 4.4b and has the following form.

(a) Select direction and assign pixel values

Select the direction to be NS and the corresponding three consecutive pixels to be a, b, c (see Fig. 4.4a and Fig. 4.4b, respectively).

(b) Carry out central pixel adjustments

Do the following intensity adjustments (see Fig. 4.4.b)

if $a \geq b + 2$ then $b = b + 1$,

if $a \succ b$ and $b \leq c$ then $b = b + 1$,

if $c \succ b$ and $b \leq a$ then $b = b + 1$,

if $c \geq b + 2$ then $b = b + 1$,

if $a \leq b - 2$ then $b = b - 1$,

if $a \prec b$ and $b \geq c$ then $b = b - 1$,

if $c \prec b$ and $b \geq a$ then $b = b - 1$,

if $c \leq b - 2$ then $b = b - 1$.

(c) Repeat

Repeat steps 1 and 2 for directions west-east (WE) direction, west-north to south-east (WN-SE), and north-east to west-south direction (NE to WS) (see Fig. 4.4a).

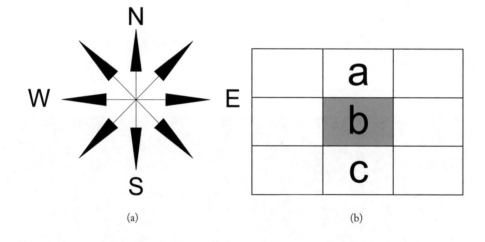

(a)   (b)

Figure 4.4: (a) Directions of implementation of the *DsFgf4d* geometric filter and (b) pixels selected for the NS direction (intensity of central pixel b is adjusted based on the values of intensities of pixels a, b for the geometric filter *DsFgf4d*).

Table 4.4: Selected statistical and image quality features for Fig. 4.5 before and after despeckle filtering for the *DsFgf4d* despeckle filter. Bold values show improvement after despeckle filtering

Features	μ	Median	σ^2	σ^3	σ^4	Contrast	C	CSR	MSE	PSNR	Q	SSIN	AD	SC	MD	NAE
Original	36	37	21	0.2	2.8	76	58	-	-	-	-	-	-	-	-	-
DsFgf4d	45	47	21	0.02	2.9	55	**47**	75	1736	26	0.62	0.7	-0.83	0.71	101	0.23

Algorithm 4.4 presents the algorithmic steps for the implementation of the *DsFgf4d* despeckle filter together with the MATLAB™ code.

Algorithm 4.4 Nonlinear Filtering: Geometric Filter (*DsFgf4d*).

1   Load the image for filtering

2   Specify the region of interest to be filtered, the moving window size and the number of iterations
     (n) the filtering is applied to the image

3   Starting from the left upper corner of the image, within the 5x5 pixel moving window rotate a 3x3
     pixel mask around the middle pixel of the window

4   Carry out pixel adjustments as explained above

5   Assign the new greyscale value of the selected position to the middle pixel

6   Repeat steps 4 and 5 for all the pixels in the image by sliding the moving window from left to right

7   Repeat steps 3 to 6 for n iterations

8   Compute the image quality evaluation metrics and the texture features for the original and the
     despeckled images

9   Display the original and despeckled images, the image quality and evaluation metrics, and the
     texture features.

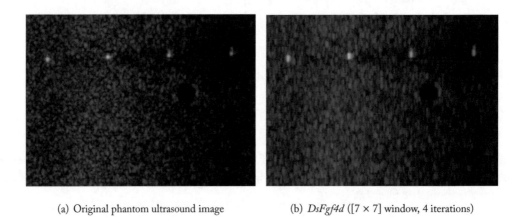

(a) Original phantom ultrasound image   (b) *DsFgf4d* ([7 × 7] window, 4 iterations)

Figure 4.5: Example of the *DsFgf4d* despeckle filtering on a phantom ultrasound image.

Figure 4.5 shows the application of the despeckle filter *DsFgf4d* on a phantom ultrasound image for a moving window size 7 × 7 pixels and 4 iterations. Table 4.4 illustrates selected statistical and image quality features for the phantom image of Fig. 4.5 for the original and the despeckled images. It is shown that the *DsFgf4d* filter increases significantly the mean and median, preserves the standard deviation and decreases the contrast C of the image.

Code 4.1 Matlab™ Code Nonlinear Filtering: Geometric Filter (*DsFgf4d*).

```
1    function f = DsFgf4d(g, nhood, niterations)
     %*************************************************************
     % Despeckle filtering toolbox, © Christos P. Loizou 2015
     % Speckle reduction filter: Dsfgf4d
     % A nonlinear geometric filter that filters the multiplicative noise in ultrasound Images. Utilizes
     % the local statistics of the noise (original) image g
     % Input Variables:
     % g :            input image to be filtered, i.e. 'cell.tif'
     % nhood :      sliding moving window, i.e [5 5]
     % niterations:  itterations for which filtering is applied iteratively
     %
     % Output Variables:
     % f: outpout image
     % Call:  f = DsFgf4d (g, [5 5], 4);
     %*************************************************************
     if isa(g, 'uint8')
       u8out = 1;
       if (islogical(g))
         logicalOut = 1;
         g = double(g);
       else
         logicalOut = 0;
         g = double(g)/255;
       end
     else
       u8out = 0;
     end
     % Specify the area of interest to be filtered, the moving window size and the number of itearions
2    % (n) the filtering is applied to the image
     % Estimate the size of the image
     [ma ,na] = size(g);
     % Estimate the midle of the processing window, which takes onle values 3, 5 7
     z=(nhood(1)-1)/2;
     %Initialize the picture f (new picture) with zeros
     f=g;

     % Starting from the left upper corner of the image, rotate a 5x5 pixel mask around the middle pixel
3    % of the window for each moving window
     for i = 1:niterations
       fprintf('\rIteration %d',i);
       if i >=2
         g=f;
       end
```

Code 4.1 *(Continued.)* Matlab™ Code Nonlinear Filtering: Geometric Filter (*DsFgf4d*).

```
% Carry out pixel adjustments as explained above (Estimate and change the middle pixel in
% the window
disp(['     Calculating/replacing the center pixel in a sliding window...']);
%ma=100; na=100;

a=1; b=0; c=3; d=1;
while c>=0,
  for d=0:1
    for i= 2 :(ma-1)
      for j= 2:(na-1)
        maxi= min(g(i-a, j-b)-1, g(i, j) +1);
        f(i, j) = max(g(i, j), maxi);
      end
      % End for j
    end
    % End for i
    for i= 2 :(ma-1)
      for j= 2:(na-1)
        maxin1 = min(f(i-a, j-b), g(i, j) +1);
        maxin = min(maxin1, f(i+a, j+b)+1);
% Assign the new greyscale value at the selected position to the middle pixel
        g(i, j) = max(f(i, j),  maxin);
      end
    % End for j
    end
    % End for i
    if d==0
      a=-a; b=-b;
    end

  end
  % End if d
  disp(['First Itteration of the Algorithm is Applied']);

  for d=0:1
    for i= 2 :(ma-1)
      for j= 2:(na-1)
        mini = max(g(i-a, j-b)+1, g(i, j) -1);
        f(i, j) = min(g(i, j), mini );
      end
    % End for j
    end
    % End for i
    for i= 2 :(ma-1)
```

4

5

5

Code 4.1 *(Continued.)* Matlab™ Code Nonlinear Filtering: Geometric Filter (*DsFgf4d*). *(Continues.)*

```
              for j= 2:(na-1)
                  mini1 = max(f(i-a, j-b), g(i, j) -1);
                  minin = max (mini1, f(i+a, j+b)-1);
                  g(i, j) = min(f(i, j),  minin);
              end
      % End for j
        end
        % End for i
        if d==0
            a=-a; b=-b;
        end
      % End if d
        end
      % End for d
        disp(['Second Itteration of the Algorithm is Applied']);
        % Repeat steps 4 and 5 for all the pixels in the image by sliding the moving window from left to
        % right
        switch c
        case 3
          a=0; b=1; c=2;
          break;
        case 2
          a=1; b=1; c=1;
          break;
        case 1
          a=1; b=-1; c=0;
          break;
        case 0
          c=-1;
          break;
        end
        % End switch
        end
        % End while c>=0 loop
        end
        % Repeat steps 3 to 6 for n iterations
        fprintf('\n');

        if u8out==1,
          if (logicalOut)
            f = uint8(f);
          else
            f = uint8(round(f*255));
```

5

6

7

Code 4.1 *(Continued.)* Matlab™ Code Nonlinear Filtering: Geometric Filter (*DsFgf4d*).

```
      end
    end
    figure, imshow(f);
    title('Image filtered by gf4d filter');
    % Compute the texture features and image quality evaluation metrics and display both the original
    % and the despeckled images on the screen
    TAM=DsTtexfeat(double(f));
    F1=[F1,TAM'];
    save speckle1texfs F1;
    % The texture features of the despeckled image are saved in matrix F1
    % Call the function metrics to calculate and display the 19 different image quality metrics between
8   % the original and the despeckled image
    M=DsQmetrics(f, imagec);

    % Show original and despeckled images on the screen
    figure, imshow(a), title ('Original Image');
    figure, imshow(f), title ('Despeckled Image');
```

4.5   HOMOMORPHIC FILTERING (*DSFHOMO*)

The *DsFhomo* filter performs homomorphic filtering for image enhancement by computing the Fast Fourier Transform (FFT) of the logarithmic compressed image, applying a denoising homomorphic filter function $H(\cdot)$, and then performing the inverse FFT of the image [51, 59]. The homomorphic filter function $H(\cdot)$, maybe constructed either using a band-pass Butterworth or a high-boost Butterworth filter. In this book, a high-boost Butterworth filter was used with the homomorphic function [51]:

$$H_{u,v} = \gamma_L + \frac{\gamma_H}{1 + (D_0/D_{u,v})^2} \qquad (4.3)$$

with

$$D_{u,v} = \sqrt{(u - N/2)^2 + (v - N/2)^2}, \qquad (4.4)$$

where $D_0 = 1.8$ is the cut of frequency of the filter, and $\gamma_L = 0.4, \gamma_H = 0.6$, are the gains for the low and high frequencies, respectively, u, v, are the spatial coordinates of the frequency transformed image, and N the dimensions of the image in the u, v, space. This form of filtering sharpens features and flattens speckle variations in an image.

Algorithm 4.5 presents the algorithmic steps for the implementation of the *DsFhomo* filter. Figure 4.6 illustrates the application of the despeckle filter *DsFhomo* on a phantom ultrasound image. Table 4.5 illustrates selected statistical and image quality features for the phantom image of Fig. 4.6 for the original and the despeckled images. It is shown that the filter increases enormously the mean, median, variance, and contrast. The Q and *SSIN* are very low.

Algorithm 4.5 Nonlinear Filtering: Homomorphic Filter (*DsFhomo*).

1	Load the image for filtering
2	Calculate the FFT of the image
3	Construct a denoising homomorphic filter using H as in (4.3 and 4.4)
4	Apply the denoising homomorphic filter function H calculated in step 3
5	Perform the inverse FFT of the image to form the despeckled image
6	Compute the image quality evaluation metrics and the texture features for the original and the despeckled images
7	Display the original and despeckled images, the image quality and evaluation metrics, and the texture features.

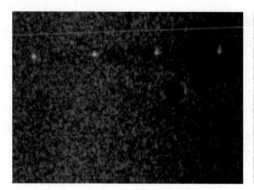

(a) Original phantom ultrasound image   (b) *DsFhomo* ([7 × 7] window, 1 iteration)

Figure 4.6: Example of *DsFhomo* despeckle filtering on a phantom ultrasound image.

4.6 HYBRID MEDIAN FILTERING (*DSFHMEDIAN*)

The filter *DsFhmedian* which was first introduced in [81], is an extension of the *DsFmedian* filter [7, 78], applied over windows of size 5 × 5. The *DsFhmedian* filter computes the median of the outputs generated by median filtering with three different windows (cross shape window, x-shape window and normal window). The moving size window for the despeckle filter *DsFhmedian* was 5 × 5 pixels, while the number of iterations applied to each image was two. The filter preserves the edges and increases the optical perception evaluation. It can thus be used to preserve and enhance edges of various organs in ultrasound images [12, 13].

Algorithm 4.6 presents the algorithmic steps for the implementation of the *DsFhmedian* despeckle filter.

Table 4.5: Selected statistical and image quality features for Fig. 4.6 before and after despeckle filtering for the *DsFhomo* despeckle filter

Features	μ	Median	σ^2	σ^3	σ^4	Contrast	C	CSR	MSE	PSNR	Q	SSIN	AD	SC	MD	NAE
Original	36	37	21	0.2	2.8	76	58	-	-	-	-	-	-	-	-	-
DsFhomo	160	176	45	-1.5	4.7	414	58	96	1760	6	0.29	0.31	-125	0.06	177	3.5

Figure 4.7 illustrates the application of the despeckle filter *DsFhmedian* on a phantom ultrasound image for a moving window size $[5 \times 5]$ pixels and 4 iterations. Table 4.6 illustrates selected statistical texture and image quality metrics features for the phantom image of Fig. 4.6 for the original and the despeckled images. It is shown that the mean and median are preserved, the variance is reduced, *PSNR*, *Q*, and *SSIN* are high.

Algorithm 4.6 Nonlinear Filtering: Median Filter (*DsFhmedian*).

---

1   Load the image for filtering

2   Specify the region of interest to be filtered and the number of iterations (n). The moving window size is set to 5.

3   Starting from the left upper corner of the image, compute the median value for three different sliding moving windows (cross shape window, x-shape window and normal window) and then compute the median for these three values

4   Replace the middle pixel in the sliding window with the median value calculated in step 3

5   Repeat steps 3 and 4 for the whole image by sliding the moving window from left to right

6   Repeat steps 3 to 5 for n iterations

7   Compute the image quality evaluation metrics and the texture features for the original and the despeckled images

8   Display the original and despeckled images, the image quality and evaluation metrics, and the texture features.

4.7   KUWAHARA FILTERING (*DSFKUWAHARA*)

The *DsFKuwahara* is an 1D filter operating in a 5×5 pixel neighborhood searching for the most homogenous neighborhood area around each pixel [13, 82]. The middle pixel of the 1×5 neighborhood is then substituted by the median gray level of the 1×5 mask (see also Fig. 4.8). The filter is iteratively applied to the image where the number of iterations was set to two. The *Ds-*

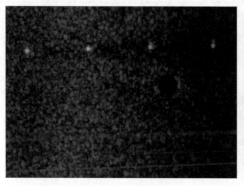 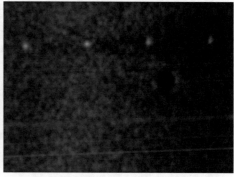

(a) Original phantom ultrasound image   (b) *DsFhmedian* ([5 × 5] window, 4 iterations)

Figure 4.7: Example of the *DsFhmedian* despeckle filtering on a phantom ultrasound image.

Table 4.6: Selected statistical and image quality features for Fig. 4.7 before and after despeckle filtering for the *DsFhmedian* despeckle filter. Bold values show improvement after despeckle filtering

Features	μ	Median	σ^2	σ^3	σ^4	Contrast	C	CSR	MSE	PSNR	Q	SSIN	AD	SC	MD	NAE
Original	36	37	21	0.2	2.8	76	58	-	-	-	-	-	-	-	-	-
DsFhmedian	36	34	18	0.08	2.4	11	**50**	1	1730	30	0.63	0.74	0.34	1.08	54	0.17

Fkuwahara filter can be used to improve the classification accuracy of different organs and tissues and to enhance edges, thus also improving the optical perception evaluation [13].

Algorithm 4.7 presents the algorithmic steps for the implementation of the *DsFKuwahara* despeckle filter.

Figure 4.8 shows the application of the despeckle filter *DsFKuwahara* on a phantom ultrasound image for a moving window size 5 × 5 pixels and 2 iterations. Table 4.7 illustrates selected statistical and image quality features for the phantom image of Fig. 4.8 for the original and the despeckled images. It is shown that the *DsFkuwahara* filter preserves mean, median and variance while contrast is slightly increased. Moreover, Q and *SSIN* are high.

Table 4.7: Selected statistical and image quality features for Fig. 4.8 before and after despeckle filtering for the *DsFkuwahara* despeckle filter

Features	μ	Median	σ^2	σ^3	σ^4	Contrast	C	CSR	MSE	PSNR	Q	SSIN	AD	SC	MD	NAE
Original	36	37	21	0.2	2.8	76	58	-	-	-	-	-	-	-	-	-
DsFkuwahara	36	37	21	0.16	2.8	79	58	15	1730	30	0.63	0.74	0.34	1.08	54	0.17

Algorithm 4.7 Nonlinear Filtering: Kuwahara Filter (*DsFKuwahara*).

---

1   Load the image for filtering

2   Specify the region of interest to be filtered and the number of iterations (n). The moving window size is set to 5.

3   Starting from the left upper corner of the image, compute for each row and each column in the 5x5 window their median values

4   Replace the middle pixel in the sliding window with the median value calculated in step 3

5   Repeat steps 3 and 4 for the whole image by sliding the moving window from left to right

6   Repeat steps 3 to 5 for n iterations

7   Compute the image quality evaluation metrics and the texture features for the original and the despeckled images

8   Display the original and despeckled images, the image quality and evaluation metrics, and the texture features.

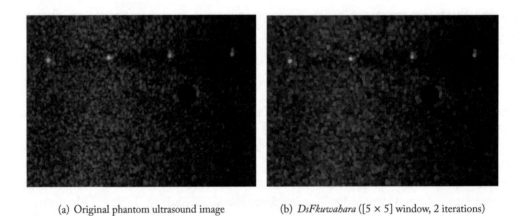

(a) Original phantom ultrasound image   (b) *DsFkuwahara* ([5 × 5] window, 2 iterations)

Figure 4.8: Example of the *DsFkuwahara* despeckle filtering on a phantom ultrasound image.

4.8   NONLOCAL FILTERING (*DSFNLOCAL*)

The *DsFnlocal* is a nonlinear 2D filter which replaces the center pixel in the moving window, with the mean of the values of all pixels in the window whose Gaussian neighborhood looks like the middle pixel in the window [83]. The main difference of the *DsFnlocal* filter with respect to local filters or frequency domain filters is the systematic use of all possible self-predictions the image can provide, according to [104] and [105]. The *DsFnlocal* filter acts as weighted average of all the pixels in the window where the family of weights $(w_{i,j})$ depend on the similarity between the pixels. The similarity between two pixels depends on the similarity of the intensity gray level vectors of those pixels which are surrounding the middle pixel in the moving window. This similarity is measured as a decreasing function of the weighted Euclidean distance between two pixels. For a more detailed analysis on the NL-means algorithm see [106].

$$f_{i,j} = \sum_{i=1}^{N}\sum_{j=1}^{M} w_{i,j} g_{i,j} \quad \text{with} \quad 0 \leq w_{i,j} \leq 1 \quad \text{and} \quad \sum_{i}\sum_{j} w_{i,j} = 1. \qquad (4.5)$$

The *DsFnlocal* filter not only compares the grey level in a single point but also the geometrical configuration in the whole neighborhood. This fact may allow a more robust comparison than other neighborhood filters. The filter is iteratively applied to the image where the number of iterations was set to two. The filter can be used to improve the classification accuracy of different organs and tissues and to enhance edges, thus also improving the optical perception evaluation [13]. It operates in a [5 × 5] up to [21 × 21] pixel neighborhood. In [105], an enhanced method of the *DsFnlocal* filtering was presented by using similarity weight calculation for adapting the filter where similar results as in [83] were presented.

Algorithm 4.8 presents the algorithmic coding steps for the implementation of the *DsFnlocal*.

Figure 4.9 shows the application of the despeckle filter *DsFnlocal* on a phantom ultrasound image for a moving window size 5 × 5 pixels and 2 iterations. Table 4.8 illustrates selected statistical and image quality features for the phantom image of Fig. 4.9 for the original and the despeckled images. It is shown that the *DsFnlocal* filter preserves the mean, reduces the standard deviation and speckle index, C, of the image.

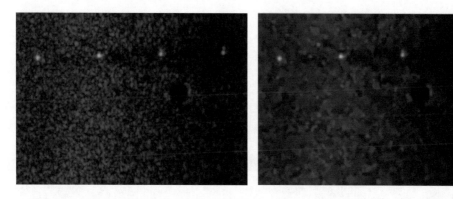

(a) Original phantom ultrasound image   (b) *DsFnlocal* ([5 × 5] window, 2 iterations)

Figure 4.9: Example of the *DsFnlocal* despeckle filtering on a phantom ultrasound image.

Algorithm 4.8 Nonlinear Filtering: Nonlocal Filter (*DsFnlocal*).

---

1   Load the image for filtering

2   Specify the region of interest to be filtered, the number of iterations (n) and the moving window size

3   Starting from the left upper corner of the image, compute for each row and each column in the moving window, the similarity for each pixel with the middle pixel in the window according to (4.5)

4   Replace the middle pixel in the sliding window with the mean grayscale value of the row or column with the highest similarity, calculated in step 3

5   Repeat steps 3 and 4 for the whole image by sliding the moving window from left to right

6   Repeat steps 3 to 5 for n iterations

7   Compute the image quality evaluation metrics and the texture features for the original and the despeckled images

8   Display the original and despeckled images, the image quality and evaluation metrics, and the texture features.

Table 4.8: Selected statistical and image quality features for Fig. 4.9 before and after despeckle filtering for the *DsFnlocal* despeckle filter. Bold values show improvement after despeckle filtering

Features	μ	Median	σ^2	σ^3	σ^4	Contrast	C	CSR	MSE	PSNR	Q	SSIN	AD	SC	MD	NAE
Original	36	37	21	0.2	2.8	76	58	-	-	-	-	-	-	-	-	-
DsFkuwahara	36	37	21	0.16	2.8	79	58	15	1730	30	0.63	0.74	0.34	1.08	54	0.17

CHAPTER 5

Diffusion Despeckle Filtering

In this chapter we present the basic theoretical background of diffusion filtering techniques together with their algorithmic implementation MATLAB™ code for selected filters and practical examples on phantom images. There are four different filters presented in this chapter which are the following: anisotropic diffusion, speckle reducing anisotropic diffusion, nonlinear anisotropic diffusion, and nonlinear coherent diffusion filtering (see also Table 1.3).

Diffusion filters remove noise from an image by modifying the image via solving a partial differential equation (PDE). The smoothing is carried out depending on the image edges and their directions. Anisotropic diffusion is an efficient nonlinear technique for simultaneously performing contrast enhancement and noise reduction. It smoothes homogeneous image regions but retains image edges [27, 29, 39] without requiring any information from the image power spectrum. It may, thus, directly be applied to images.

Consider applying the isotropic diffusion equation given by $d g_{i,j,t}/dt = div(d\nabla g)$ using the original noisy image, $g_{i,j,t=0}$, as the initial condition, where $g_{i,j,t=0}$ is an image in the continuous domain, where i, j, specifies spatial position, t, is an artificial time parameter, d, is the diffusion constant, and ∇g, is the image gradient. Modifying the image according to this linear isotropic diffusion equation is equivalent to filtering the image with a Gaussian filter. In this chapter we will present conventional anisotropic diffusion (*DsFad*), speckle reducing anisotropic diffusion (*DsFsrad*), coherent nonlinear anisotropic diffusion (*DsFnldif*), and nonlinear complex diffusion (*DsFncdif*).

5.1   ANISOTROPIC DIFFUSION FILTERING (*DSFAD*)

Perona and Malik [24] replaced the classical isotropic diffusion equation, as described above, by the introduction of a function, $d_{i,j,t} = f(|\nabla g|)$, that smoothes the original image while trying to preserve brightness discontinuities with:

$$\frac{d g_{i,j,t}}{dt} = div\left[d_{i,j,t}\nabla g_{i,j,t}\right] = \left[\frac{d}{di}d_{i,j,t}\frac{d}{di}g_{i,j,t}\right] + \left[\frac{d}{dj}d_{i,j,t}\frac{d}{dj}g_{i,j,t}\right], \qquad (5.1)$$

where $|\nabla g|$ is the gradient magnitude, and $d(|\nabla g|)$, is an edge stopping function, which is chosen to satisfy $d \to 0$ when $|\nabla g| \to \infty$ so that the diffusion is stopped across edges. This function, called the diffusion coefficient $d(|\nabla g|)$, which is a monotonically decreasing function of the gradient magnitude, $|\nabla g|$, yields intra-region smoothing, and not inter-region smoothing [27, 39, 60, 61], by impeding diffusion at image edges. It increases smoothing parallel to the edge and stops smoothing perpendicular to the edge, as the highest gradient values are perpendicular to the edge and dilated across edges. The choice of $d(|\nabla g|)$, can greatly affect the extent to which discontinuities are preserved. For example, if $d(|\nabla g|)$, is constant at all locations, then smoothing progresses in an isotropic manner. If $d(|\nabla g|)$, is allowed to vary according to the local image gradient, then we have anisotropic diffusion. A basic anisotropic PDE is given in (5.1). Two different diffusion coefficients were proposed in [27] and also derived in [39]. The diffusion coefficient suggested were:

$$d(|\nabla g|) = \frac{1}{1 + \left(|\nabla g_{i,j}|/K\right)^2}, \quad \text{and} \quad cd(|\nabla g|) = \frac{2\,|\nabla g_{i,j}|}{2 + \left(|\nabla g_{i,j}|/K_1\right)^2}, \tag{5.2}$$

where K and K_1 in (5.2) are a positive gradient threshold parameter, known as diffusion or flow constant [39]. In this book the first diffusion coefficient in (5.2) was used as it was found to perform better in our images [3, 7].

A discrete formulation of the anisotropic diffusion in (5.1) is [27, 32, 39]:

$$\frac{dg_{i,j}}{dt} = \frac{\lambda}{|\eta_s|} \Big\{ d_{i+1,j,t} \left[g_{i+1,j} - g_{i,j}\right] + d_{i-1,j,t} \left[g_{i-1,j} - g_{i,j}\right] \\ + d_{i,j+1,t} \left[g_{i,j+1} - g_{i,j}\right] + d_{i,j-1,t} \left[g_{i,j-1} - g_{i,j}\right] \Big\}, \tag{5.3}$$

where the new pixel gray value, $f_{i,j}$, at location i, j, is

$$f_{i,j} = g_{i,j} + \frac{1}{4} \frac{dg_{i,j}}{dt}, \tag{5.4}$$

where $d_{i+1,j,t}$, $d_{i-1,j,t}$, $d_{i,j+1,t}$, and $d_{i,j-1,t}$, are the diffusion coefficients for the west, east, north, and south pixel directions, in a four pixel neighborhood, around the pixel i, j, where diffusion is computed respectively. The diffusion coefficient leads to the largest diffusion where the nearest-neighbor difference is largest (largest edge), while the smallest diffusion is calculated where the nearest-neighbor difference is smallest (the weakest edge). The constant, $\lambda \in \Re^+$, is a scalar that determines the rate of diffusion, η_s, represents the spatial neighborhood of pixel, i, j, and $|\eta_s|$, is the number of neighbors (usually four except at the image boundaries). Perona and Malik [27] linearly approximated the directional derivative in a particular direction as, $\nabla g_{i,j} = g_{i+1,j} - g_{i,j}$, (for the east direction of the central pixel i, j). Modifying the image according to the above equation in (5.3), which is a linear isotropic diffusion equation, is equivalent to filtering the image with a Gaussian filter. The parameters for the anisotropic diffusion filter used in this book were, $\lambda = 0.25$, $\eta_s = 8$, and the parameter $K = 30$, which was used for the calculation of the edge stopping function $d(|\nabla g|)$, in (5.2).

Figure 5.1 shows the application of the despeckle filter *DsFad* on a phantom ultrasound image for 40 iterations. Table 5.1 illustrates selected statistical and image quality features for the phantom image of Fig. 5.1 for the original and the despeckled images. It is shown that the *DsFad* filter preserves the mean and median, reduces the standard deviation, contrast and speckle index, C, of the image. The *PSNR*, Q, and *SSIN* are high while AD and MD are low.

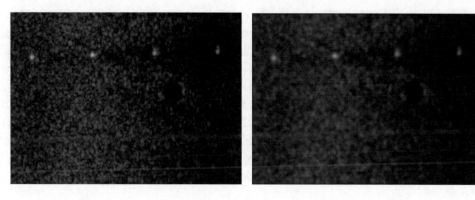

(a) Original phantom ultrasound image   (b) *DsFad* (40 iterations)

Figure 5.1: Example of the *DsFad* despeckle filtering on a phantom ultrasound image.

Table 5.1: Selected statistical and image quality features for Fig. 5.1 before and after despeckle filtering for the *DsFad* despeckle filter. Bold values show improvement after despeckle filtering

Features	μ	Median	σ^2	σ^3	σ^4	Contrast	C	CSR	MSE	PSNR	Q	SSIN	AD	SC	MD	NAE
Original	36	37	21	0.2	2.8	76	58	-	-	-	-	-	-	-	-	-
DsFad	36	38	**18**	0.03	2.5	15	48	1	1741	31	0.71	0.78	0.33	1.08	58	0.16

5.2   SPECKLE-REDUCING ANISOTROPIC DIFFUSION FILTERING (*DSFSRAD*)

The essence of speckle reducing anisotropic diffusion is the replacement of the gradient based edge detector, $cd(|\nabla g|)$ in original anisotropic diffusion PDE with the instantaneous coefficient of variation suitable for speckle filtering, $c_{srad}(|\nabla g|)$. The *DsFsrad* speckle reducing anisotropic diffusion filter [29], uses two seemingly different methods, namely the Lee [36, 38, 48] and the Frost diffusion filters [47]. A more general updated function for the output image by extending the PDE versions of the despeckle filter is [29]:

$$f_{i,j} = g_{i,j} + \frac{1}{\eta_s} div \left(c_{srad} \left(|\nabla g| \right) \nabla g_{i,j} \right). \tag{5.5}$$

The diffusion coefficient for the speckle anisotropic diffusion, $c_{srad}(|\nabla g|)$, is derived [29]as:

$$c^2_{srad} \left(|\nabla g| \right) = \frac{\frac{1}{2}|\nabla g_{i,j}|^2 - \frac{1}{16}(\nabla^2 g_{i,j})^2}{\left(g_{i,j} + \frac{1}{4}\nabla^2 g_{i,j} \right)^2}. \tag{5.6}$$

It is required that $c_{srad}(|\nabla g|) \geq 0$. The above instantaneous coefficient of variation combines a normalized gradient magnitude operator and a normalized Laplacian operator to act like an edge detector for speckle images. High relative gradient magnitude and low relative Laplacian indicates an edge. The *DsFsrad* filter utilizes speckle reducing anisotropic diffusion after (5.4) with the diffusion coefficient, $c_{srad}(|\nabla g|)$ in (5.6) [29]

Algorithm 5.1 presents the algorithmic steps for the implementation of the *DsFsrad* filter together with the MATLAB™ code. Figure 5.2 shows the results of the *DsFrad* filter on a phantom image, with 40 iterations. Table 5.2 illustrates selected statistical and image quality features for the phantom image of Fig. 5.2 for the original and the despeckled images. It is shown that the *DsFsrad* filter does not preserve the statistical measures after filtering, while the contrast and C are reduced.

Algorithm 5.1 Diffusion Filtering: Speckle Reducing Anisotropic Filter (*DsFsrad*).

1	Load the image for filtering
2	Specify the original image to be filtered (K), the number of iterations (n), the time step (lamda), and the region of interest to be filtered (rect)
3	Transform the original image to double and normalize it to: $f_{i,j} = (f_{i,j} - \min pixelvalue)/(\max pixelvalue - \min pixelvalue)$, where minpixelvalue, and maxpixelvalue represents the minimum and maximum pixel values in the image.
4	Starting from the left upper corner of the image, select a 3x3 pixel neighbourhood and compute a new greyscale value according to Eq. (5.5)
5	Assign the new greyscale value to the middle pixel in each window
6	Repeat steps 4 and 5 for the whole image by sliding the moving window from left to right
7	Repeat steps 4 to 6 for niter iterations
8	Compute the image quality evaluation metrics and the texture features for the original and the despeckled images
9	Display the original and despeckled images, the image quality and evaluation metrics, and the texture features.

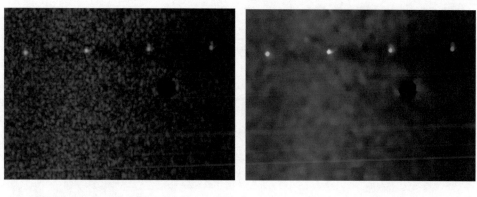

(a) Original phantom ultrasound image   (b) *DsFsrad* (40 iterations)

Figure 5.2: Example of the *DsFsrad* despeckle filtering on a phantom ultrasound image.

Table 5.2: Selected statistical and image quality features for Fig. 5.2 before and after despeckle filtering for the *DsFsrad* despeckle filter. Bold values show improvement after despeckle filtering

Features	μ	Median	σ^2	σ^3	σ^4	Contrast	C	CSR	MSE	PSNR	Q	SSIN	AD	SC	MD	NAE
Original	36	37	21	0.2	2.8	76	58	-	-	-	-	-	-	-	-	-
DsFsrad	50	56	24	0.16	2.8	17	**48**	11	1755	23	0.18	0.45	-14	0.57	83	0.43

5.3 NONLINEAR ANISOTROPIC DIFFUSION FILTERING (*DSFNLDIF*)

The applicability of the *DsFad* filter (5.5) is restricted to smoothing with edge enhancement, where $|\nabla g|$, has higher magnitude at edges. In general, the function $d\left(|\nabla g|\right)$, in (5.5) can be put into a tensor form that measures local coherence of structures such that the diffusion process becomes more directional in both the gradient and the contour directions, which represent the directions of maximum and minimum variations, respectively. Therefore, the *DsFnldif* filter will take the form:

$$\frac{d g_{i,j,t}}{dt} = div\left[D\nabla g\right], \tag{5.7}$$

where $D \in \Re^{2\times2}$ is a symmetric positive semi-definite diffusion tensor representing the required diffusion in both gradient and contour directions and, hence, enhancing coherent structures as well as edges. The design of D, as well as the derivation of the coherent nonlinear anisotropic

Code 5.1 Matlab™ Code Diffusion Filtering: Speckle Reducing Anisotropic Filter (*DsFsrad*) (*Continues.*)

```
1    function [I,rect] = DsFSRAD(K,niter,lambda,rect)
     %**********************************************************************
     % Load the image for filtering
2    % Specify the area of interest to be filtered (I), the number of iterations (niter), the time step
     % (lamda), and the area to be filtered (rect)
     % Despeckle filtering toolbox, % © Christos P. Loizou 2015
     % Speckle filtering using SRAD (Speckle Reducing Anisotropic Diffusion)
     % Input Variables:
     % K: original image
     % niter = number of iterations to apply the filter
     % lambda = time step
     % rect: rectangle area to be filtered
     %
     % Output Variables:
     % I = new smoothed image
     % rect = region of interest (ROI)
     %
     % Example 1:  [I,rect] = DsFSRAD(K(:,:,1),75,0.025);
     % Example 2:  [I,rect] = DsFSRAD(K(:,:,1),75,0.025, [0 0 436 182]);
     % to despeckle the image by directly defining the ROI to be filtered
     %**********************************************************************
3        % Transform the original image to double and normalize it
     I = double(K);
     mx = max(I(:));
     mn = min(I(:));
     I = (I-mn)/(mx-mn);
     % indices (using boudary conditions)
     [M,N] = size(I);
     iN = [1, 1:M-1];
     iS = [2:M, M];
     jW = [1, 1:N-1];
     jE = [2:N, N];
     % get an area of uniform speckle
     if nargin < 4 || isempty(rect)
         imshow(I,[],'notruesize');
         rect = getrect;
     end
     % log uncompress the image and eliminate zero value pixels.
     I = exp(I);
     % wait bar
     hwait = waitbar(0,'Diffusing Image');
```

Code 5.1 *(Continued.)* Matlab™ Code Diffusion Filtering: Speckle Reducing Anisotropic Filter (*DsFsrad*) *(Continues.)*

| 4 | ```
% Starting from the left upper corner of the image, select a 3x3 pixel neighbourhood and compute a
% new greyscale value according to (5.5)
for iter = 1:niter
    % speckle scale function
    Iuniform = imcrop(I,rect);
    q0_squared = (std(Iuniform(:))/mean(Iuniform(:)))^2;

    % differences
    dN = I(iN,:) - I;
    dS = I(iS,:) - I;
    dW = I(:,jW) - I;
    dE = I(:,jE) - I;
     % normalized discrete gradient magnitude squared
    G2 = (dN.^2 + dS.^2 + dW.^2 + dE.^2) ./ (I.^2 + eps);
    % normalized discrete laplacian
    L = (dN + dS + dW + dE) ./ (I + eps);
     % ICOV (equ 31/35)
    num = (.5*G2) - ((1/16)*(L.^2));
    den = (1 + ((1/4)*L)).^2;
    q_squared = num ./ (den + eps);
     % diffusion coefficent
    den = (q_squared - q0_squared) ./ (q0_squared *(1 + q0_squared) + eps);
    c = 1 ./ (1 + den);
    cS = c(iS, :);
    cE = c(:,jE);
     % divergence
``` |
| 5 | ```
    D = (cS.*dS) + (c.*dN) + (cE.*dE) + (c.*dW);
     % Assign the new greyscale value to the middle pixel in each window
    I = I + (lambda/4)*D;
``` |
| 6 | ```
% Repeat steps 4 and 5 for the whole image by sliding the moving window from left to right
    waitbar(iter/niter,hwait);
end
``` |
| 7 | ```
% Repeat steps 4 to 6 for niter iterations
I = log(I);
figure, imshow(I);
% close wait bar
close(hwait)
return;
``` |

Code 5.1 *(Continued.)* Matlab™ Code Diffusion Filtering: Speckle Reducing Anisotropic Filter (*DsFsrad*)

```
8    % Compute the texture features and image quality evaluation metrics and display both the original
     % and the despeckled image on the screen
     TAM=DsTtexfeat(double(I));
     F1=[F1,TAM'];
     save speckle1texfs F1;
     % The texture features of the despeckled image are saved in matrix F1
     % Call the function metrics to calculate and display the 19 different image quality metrics between
     %  the original and the despeckled image
     M=DsQmetrics(I, K);

     % Show original and despeckled images on the screen
     figure, imshow(K), title ('Original Image');
     figure, imshow(I), title ('Despeckled Image');
```

diffusion model may be found in [3] and is given as:

$$D = (\omega_1 \ \omega_2) \begin{pmatrix} \lambda_1 & 0 \\ 0 & \lambda_2 \end{pmatrix} \begin{pmatrix} \omega_1^T \\ \omega_2^T \end{pmatrix} \tag{5.8}$$

with

$$\lambda_1 = \begin{cases} \alpha \left(1 - \dfrac{(\mu_1 - \mu_2)^2}{s^2}\right) & \text{if} \quad (\lambda_1 - \lambda_2)^2 \leq s^2 \\ 0, & \text{else} \end{cases} , \tag{5.9}$$

$$\lambda_2 = \alpha,$$

where the eigenvectors ω_1, ω_2 and the eigenvalues λ_1, λ_2 correspond to the directions of maximum and minimum variations and the strength of these variations, respectively. The flow at each point is affected by the local coherence, which is measured by $(\mu_1 - \mu_2)$ in (5.9).

The parameters used in this book for the *DsFnldif* filter were, $s^2 = 2$, and $\alpha = 0.9$, which were used for the calculation of the diffusion tensor D, and the parameter step size $m = 0.2$, which defined the number of diffusion steps performed. The local coherence is close to zero in very noisy regions and diffusion becomes isotropic ($\mu_1 = \mu_2 = \alpha = 0.9$), whereas in regions with lower speckle noise the local coherence corresponds to $(\mu_1 - \mu_2)^2 > s^2$ [3].

Algorithm 5.2 presents the algorithmic steps for the implementation of the *DsFnldif* despeckle filter. Figure 5.3 illustrates the results of the *DsFnldif* filter on a phantom ultrasound image, whereas Table 5.3 shows statistical and image quality features before and after despeckle filtering. It is shown that mean, median, variance and C remain the same after despeckle filtering. The value of the quality metric *SSIN* is high.

Algorithm 5.2 Diffusion Filtering: Coherent Non Linear Anisotropic Diffusion Filter (*DsFnldif*).

---

1   Load the image for filtering

2   Specify the area of interest to be filtered, the number of iterations (n), and the time step

3   Transform the image to double and normalize it

4   Starting from the left upper corner of the image, select a 3x3 pixel neighbourhood and compute a new greyscale value according to (5.8) and (5.9)

5   Assign the new greyscale value to the middle pixel in each window

6   Repeat steps 4 and 5 for n iterations by sliding the moving window from left to right

7   Repeat steps 4 to 6 for the whole image

8   Compute the image quality evaluation metrics and the texture features for the original and the despeckled images

9   Display the original and despeckled images, the image quality and evaluation metrics, and the texture features.

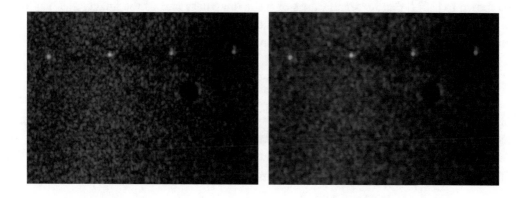

(a) Original phantom ultrasound image   (b) *DsFnldif* (30 iterations)

Figure 5.3: Example of the *DsFnldif* despeckle filtering on a phantom ultrasound image.

Table 5.3: Selected statistical and image quality features for Fig. 5.3 before and after despeckle filtering for the *DsFnldif* despeckle filter

Features	μ	Median	σ^2	σ^3	σ^4	Contrast	C	CSR	MSE	PSNR	Q	SSIN	AD	SC	MD	NAE
Original	36	37	21	0.2	2.8	76	58	-	-	-	-	-	-	-	-	-
DsFnldif	36	37	21	1.3	12	70	58	10	1746	26	0.70	0.78	-0.02	1.02	135	0.17

5.4   NONLINEAR COMPLEX DIFFUSION FILTERING (*DSFNCDIF*)

The general anisotropic complex diffusion process [3, 69] searches for the solution of (5.1), which was discretized in a finite difference scheme according to (5.2). For the *DsFncdif* filtering a difference diffusion coefficient is used in (2.12) which is derived as:

$$c_{ncdif}(|\nabla g|) = \frac{e^{i\theta}}{1 + (g_{i,j}/k\theta)^2},$$
(5.10)

where $i = \sqrt{-1}$, k a threshold parameter and θ a phase angle close to zero. In the above formulation, the definition c_{ncdif} does not involve the derivatives of the image, which is the main advantage of this approach in comparison with the *DsFad* filtering. This is because at the early stages the computations of derivatives is highly inaccurate due to the presence of noise [107]. It was shown in [107], that for small values of θ the imaginary part of the image, $g_{i,j}$, is a smooth function of its second derivatives and the ration $g_{i,j}/\theta$, is proportional to the laplacian of the image. In this way, the diffusion coefficient can be approximated by:

$$c_{ncdif}(|\nabla g|) = \frac{1}{1 + (\Delta g_{i,j}/k)^2}.$$
(5.11)

The choice of the parameter k, in (5.11) is important and can be expressed using a Lorentzian function, into a family of curves. The parameter k, modulates the spread of the diffusion coefficient in the vicinity of its maximum, that is, at edges and homogeneous areas, where the image Laplacian vanishes, and was proposed as follows [69]:

$$k = k_{max} + (k_{min} - k_{max}) \frac{g_{i,j} - \min(g_{i,j})}{\max(g_{i,j}) - \min(g_{i,j})},$$
(5.12)

where $\max(g_{i,j})$, $\min(g_{i,j})$, stand for the minimum and maximum of the image $g_{i,j}$, respectively, and k_{min}, k_{min} can be chosen from 0–10.0.

Algorithm 5.3 presents the algorithmic steps for the implementation of the *DsFncdif* despeckle filter. Figure 5.4 shows the results of the *DsFncdif* filter on an artificial carotid image for 2 iterations of the filter, while Table 5.4 illustrates the statistical and image quality features before and after despeckle filtering with where no significant changes are observed for all features after despeckle filtering. It is shown that the mean, median, and variance are preserved while *SSIN* is high. The speckle index C is increased.

Algorithm 5.3 Nonlinear Complex Diffusion Filter (*DsFncdif*).

1	Load the image for filtering
2	Specify the area of interest to be filtered, the number of iterations (n), and the time step
3	Transform the image to double and normalize it
4	Starting from the left upper corner of the image, select a 3x3 pixel neighbourhood and compute a new greyscale value according to (5.11) and (5.12)
5	Assign the new greyscale value to the middle pixel in each window according to (5.5)
6	Repeat steps 4 and 5 for n iterations by sliding the moving window from left to right
7	Repeat steps 4 to 6 for the whole image
8	Compute the image quality evaluation metrics and the texture features for the original and the despeckled images
9	Display the original and despeckled images, the image quality and evaluation metrics, and the texture features.

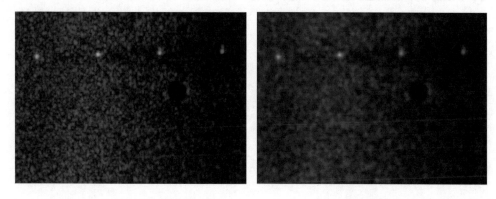

(a) Original phantom ultrasound image   (b) *DsFncdif* (30 iterations)

Figure 5.4: Example of the *DsFncdif* despeckle filtering on a phantom ultrasound image.

Table 5.4: Selected statistical and image quality features for Fig. 5.4 before and after despeckle filtering for the *DsFncdif* despeckle filter

Features	μ	Median	σ^2	σ^3	σ^4	Contrast	C	CSR	MSE	PSNR	Q	SSIN	AD	SC	MD	NAE
Original	36	37	21	0.2	2.8	76	58	-	-	-	-	-	-	-	-	-
DsFncdiff	36	38	20	1.3	13	70	71	1	1760	30	0.64	0.74	0.28	1.05	135	0.18

CHAPTER 6

Wavelet Despeckle Filtering

Wavelet filtering exploits the decomposition of the image into the wavelet basis and zeros-out the wavelet coefficients in order to despeckle the image [70, 71, 89].

Wavelet analysis is particularly useful for the analysis of transient, non-stationary, or time-varying signals. Wavelets can be used to analyse signals in different spatial resolutions. Their advantage is in their ability to analyse a signal with accuracy in both the time and frequency domains. This is not the case when applying traditional Fourier analysis, where there is significant accuracy in the frequency domain, but less accuracy in the temporal domain. In other words, increasing accuracy in one domain implies a decrease in precision in the other domain. Wavelets are also known for their capacity to identify singularities, associated with fine variations of the signal to be evaluated [71]. For denoising, we need to identify the specific image scales where the most of the image energy lies.

Speckle reduction filtering in the wavelet domain, is based on the idea of the Daubenchies Symlet wavelet and on soft-thresholding denoising. It was first proposed by Donoho [71] and also further investigated by [5, 70, 89]. The Symmlets family of wavelets, although not perfectly symmetrical, were designed to have the least asymmetry and highest number of vanishing moments for a given compact support [71]. The *DsFwaveltc* filter, implemented in this study, is described as follows.

(a) Estimate the variance of the speckle noise, σ_n^2, with (3.5).

(b) Compute the discrete wavelet transform (DWT), using the Symlet wavelet for two scales.

(c) For each sub-band:

    (a) compute a threshold [52, 71]

$$T = \begin{cases} (T_{\max} - \alpha(j-1))\sigma_n & \text{if} \quad T_{\max} - \alpha(j-1) \geq T_{\min} \\ T_{\min}\sigma_n, & \text{else,} \end{cases} \quad (6.1)$$

    where α, is a decreasing factor between two consecutive levels, $T_{\max}$, is a maximum factor for σ_n, while $T_{\min}$, is a minimum factor. The threshold T, is primarily calculated using, σ_n, and a decreasing factor, $T_{\max} - \alpha(j-1)$; and

    (b) apply the threshold on the wavelet coefficients of each band.

(d) Compute the inverse discrete wavelet transform to reconstruct the despeckled image, f.

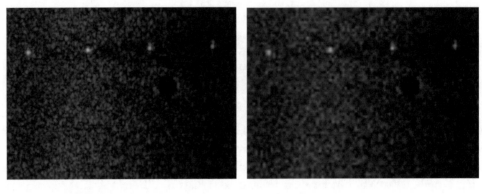

(a) Original phantom ultrasound image   (b) *DsFwaveltc* (5 iterations)

Figure 6.1: Example of the *DsFwaveltc* despeckle filtering on a phantom ultrasound image.

Table 6.1: Selected statistical and image quality features for Fig. 6.1 before and after despeckle filtering for the *DsFwaveltc* despeckle filter

Features	μ	Median	σ^2	σ^3	σ^4	Contrast	C	CSR	MSE	PSNR	Q	SSIN	AD	SC	MD	NAE
Original	36	37	21	0.2	2.8	76	58	-	-	-	-	-	-	-	-	-
DsFwaveltc	36	38	20	0.09	2.7	27	56	20	1762	31	0.66	0.72	-0.01	1.04	50	0.17

Algorithm 6.1 illustrates the algorithmic coding steps for the implementation of the *DsFwaveltc* despeckle filter. Figure 6.1 shows the application of the despeckle filter *DsFwaveltc* on a phantom ultrasound image for five iterations, whereas Table 6.1 illustrates selected statistical and image quality features for the phantom image of Fig. 6.1 for the original and the despeckled images. It is shown that the *DsFwaveltc* filter preserves the mean, median, and variance but lowers skewness and contrast of the image. The values for Q and *SSIN* are high.

Algorithm 6.1 Wavelet Filtering: Wavelet Filter (*DsFwaveltc*).

---

1   Load the image for filtering

2   Specify the area of interest to be filtered, the moving window size and the number of iterations (n) the filtering is applied to the image

3   Compute the noise variance σ_n^2 with (3.1.5) from the whole image

4   Compute the discrete wavelet transform (DWT), using the Symlet wavelet for two scales

5   Compute, for each sub band a threshold T according to (6.1)

6   Apply the threshold on the wavelet coefficients for each band

7   Compute the inverse discrete wavelet transform to reconstruct the whole image

8   Repeat steps 3 to 6 for n iterations for the whole image

9   Compute the image quality evaluation metrics and the texture features for the original and the despeckled images

10   Display the original and despeckled images, the image quality and evaluation metrics, and the texture features.

CHAPTER 7

Evaluation of Despeckle Filtering

In this chapter we present an evaluation and comparison of the 16 despeckle filtering algorithms presented in Chapters 3–6. The evaluation is carried out on a phantom image, an artificial image and on real carotid and cardiac ultrasound images. Furthermore, findings on video despeckling are presented.

7.1 DESPECKLE FILTERING EVALUATION ON AN ARTIFICIAL CAROTID ARTERY IMAGE

Despeckle filtering was evaluated on an artificial carotid artery image (see Fig. 2.1b) corrupted by speckle noise (see Fig. 7.1a). Figure 7.1 shows the original noisy image of the artificial carotid artery, degraded by speckle noise, together with the despeckled images. Figure 7.2 shows line profiles (intensity), for the line marked in Fig. 7.1a for all despeckle filters. The profile results show that most of the filters (*DsFlsmv*, *DsFwiener*, *DsFlsminsc*, *DsFmedian DsFls*, *DsFhomog*, *DsFgf4d*, *DsFhomo*, *DsFhmedian*, *DsFkuwahara*, *DsFnlocal*, *DsFad*, *DsFsrad*, *DsFnldif*, *DsFncdif* and *DsFwaveltc*). Best results were obtained for the filters *DsFmedian*, *DsFwiener*, *DsFlsmv*, *DsFhmedian*, *DsFlsminsc*, and *DsFgf4d* that preserved the edge boundaries preserving the locality and minimally affecting the reference values in each region. The filters *DsFad*, *DsFnldif*, *DsFls*, *DsFwaveltc*, *DsFhomog*, *DsFnlocal*, and *DsFhomo* do not preserve the edges, moving the line profiles to darker grayscale values. Moreover, it is shown from Fig. 7.1i that the filter *DsFhomo* is very noisy.

The despeckled images of Fig. 7.1 were also assessed by two experts. Filters that showed an improved smoothing after filtering, as assessed by visual perception criteria, were in the following order: *DsFwaveltc*, *DsFlsmv*, *DsFhmedian*, *DsFkuwahara*, *DsFlsminsc*, *DsFad*, *DsFgf4d*, and *DsFmedian*. Filters that showed a blurring effect especially on the edges were: *DsFls* and *DsFhomo*.

The upper part of Table 7.1 tabulates the statistical features μ, median, σ^2, σ^3, σ^4, the contrast, the speckle index, C (2.9), and the contrast-speckle-ratio, CSR (2.10), for the artificial image and the 16 filters illustrated in Fig. 7.1. The filters are categorized in linear filtering, nonlinear filtering, diffusion filtering, and wavelet filtering, as introduced in Chapters 3–6, respectively. Also, the number of iterations (No. of Iterations), for each despeckle filter is given, which was

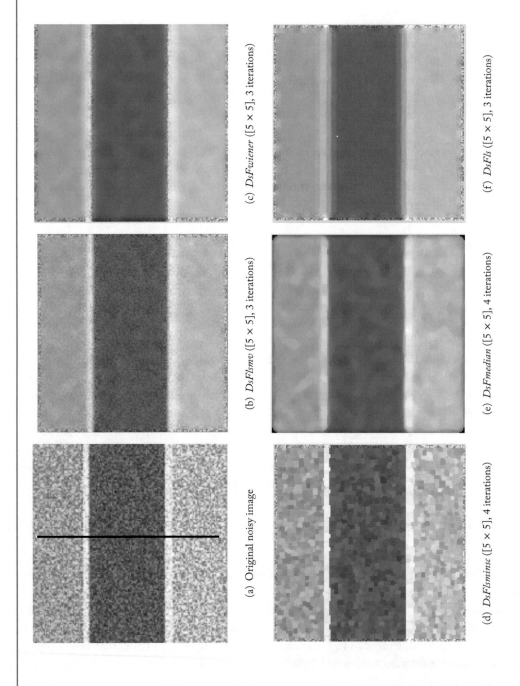

(a) Original noisy image

(b) $DsFlsmv$ ([5 × 5], 3 iterations)

(c) $DsFwiener$ ([5 × 5], 3 iterations)

(d) $DsFlsminsc$ ([5 × 5], 4 iterations)

(e) $DsFmedian$ ([5 × 5], 4 iterations)

(f) $DsFls$ ([5 × 5], 3 iterations)

Figure 7.1: Original noisy image of an artificial carotic artery given in (a), and the application of the 15 despeckle filters given in (b)–(q). (Vertical line given in (a) defines the position of the line intensity profiles plotted in Fig. 7.3) (*Continues.*)

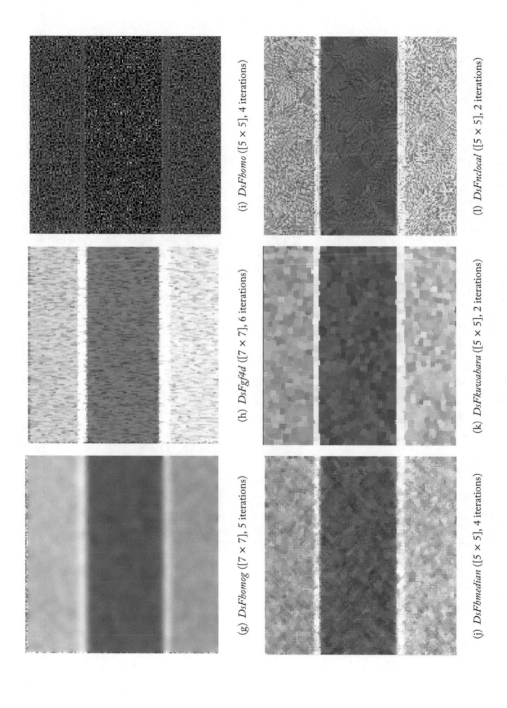

(g) *DsFhomog* ([7 × 7], 5 iterations)

(h) *DsFgf4d* ([7 × 7], 6 iterations)

(i) *DsFhomo* ([5 × 5], 4 iterations)

(j) *DsFhmedian* ([5 × 5], 4 iterations)

(k) *DsFkuwahara* ([5 × 5], 2 iterations)

(l) *DsFnclocal* ([5 × 5], 2 iterations)

Figure 7.1:   *(Continued.)* Original noisy image of an artificial carotic artery given in (a), and the application of the 15 despeckle filters given in (b)–(q). (Vertical line given in (a) defines the position of the line intensity profiles plotted in Fig. 7.3) *(Continues.)*

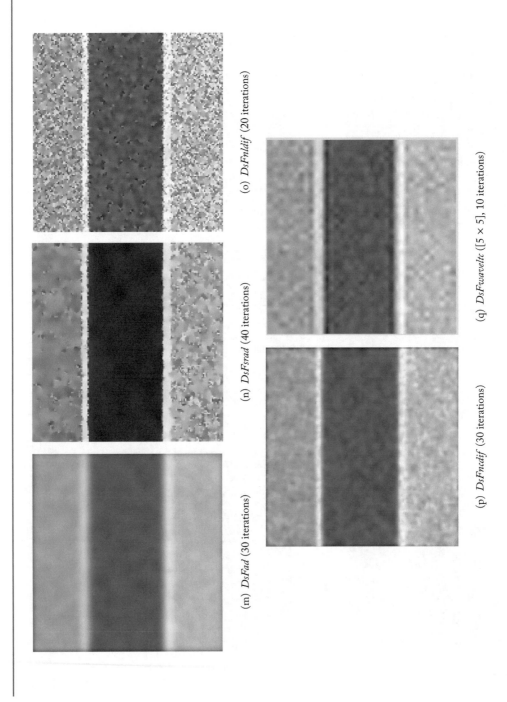

(m) *DsFad* (30 iterations)

(n) *DsFsrad* (40 iterations)

(o) *DsFnldif* (20 iterations)

(p) *DsFncdif* (30 iterations)

(q) *DsFwaveltc* ([5 × 5], 10 iterations)

Figure 7.1: *(Continued.)* Original noisy image of an artificial carotid artery given in (a), and the application of the 15 despeckle filters given in (b)–(q). (Vertical line given in (a) defines the position of the line intensity profiles plotted in Fig. 7.3)

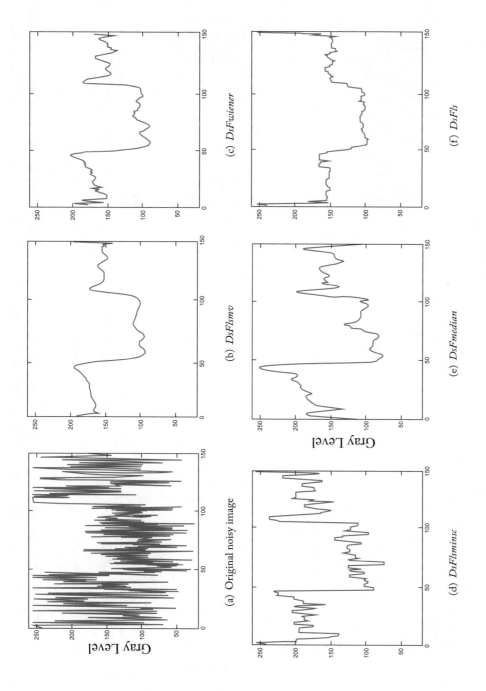

Figure 7.2:   Line profiles of the line illustrated in Fig. 7.1a for the original noisy image (a), and the 16 despeckled images given in (b)–(q). (a) Original phantom carotid image, (b) *DsFlsmv*, (c) *DsFwiener*, (d) *DsFlsminsc*, (e) *DsFmedian*, and (f) *DsFls*. (*Continues.*)

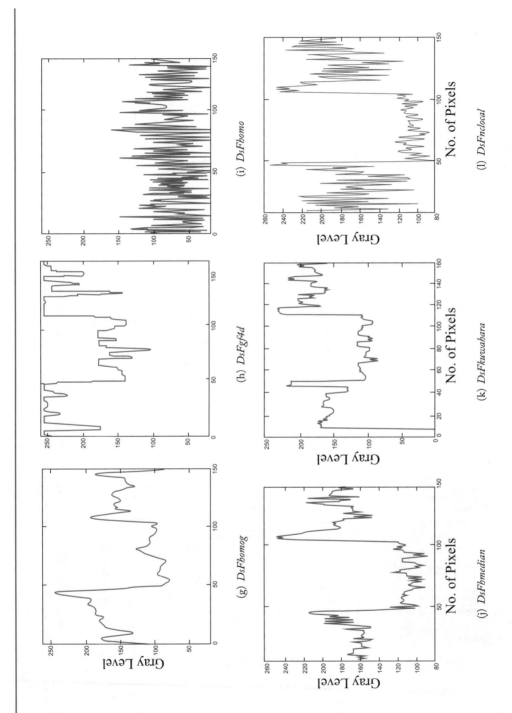

Figure 7.2: *(Continued.)* Line profiles of the line illustrated in Fig. 7.1a for the original noisy image (a), and the 16 despeckled images given in (b)–(q). (g) *DsFhomog*, (h) *DsFgf4d*, (i) *DsFhomo*, (j) *DsFmedian*, (k) *DsFKuwahara*, and (l) *DsFnlocal*. *(Continues.)*

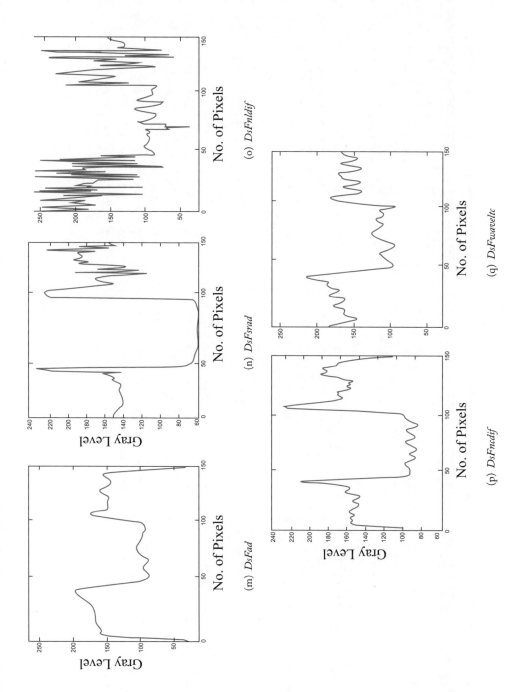

Figure 7.2: (*Continued.*) Line profiles of the line illustrated in Fig. 7.1a for the original noisy image (a), and the 16 despeckled images given in (b)–(q). (m) *DsFad*, (n) *DsFsrad*, (o) *DsFnldif*, (p) *DsFncdif*, and (q) *DsFwaveltc*.

Table 7.1: Selected statistical and image quality evaluation metrics for images of Fig. 7.1 before and after despeckle filtering

Feature	original image	Linear Filtering						Non-Linear Filtering					Diffusion				Wavelet
		DsF lsmv	DsF lsminsc	DsF wiener	DsF median	DsF ls	DsF homog	DsF gf4d	DsF homo	DsF hmedian	DsF Kuwahara	DsF nlocal	DsF ad	DsF srad	DsF nldif	DsF ncdif	DsF waveltc
No. of iterations		2	1	2	2	3	3	4	2	4	2	2	20	50	5	20	5
Window size		5x5	5x5	5x5	5x5	5x5	5x5	5x5	5x5	5x5	5x1	5x5	-	-	-	-	5x5
μ	138	145	157	145	145	143	145	176	55	146	142	146	139	145	143	145	146
Median	132	151	162	157	152	157	156	157	55	151	150	135	152	135	132	155	156
σ^2	53	41	46	37	40	33	40	46	24	41	50	47	39	55	51	39	38
σ^3	0.85	-0.1	0.09	-0.2	0.07	-0.2	0.02	0.07	0.36	0.20	-0.37	0.51	-0.35	0.75	0.44	0.06	-0.09
σ^4	2.0	2.0	1.8	1.6	2.0	1.8	1.8	1.8	4	2.0	3.4	2.1	2	2.2	2	1.8	1.6
Contrast	124	68	239	27	141	201	132	1072	340	155	207	695	26	54	60	29	50
C	38	28	29	26	28	23	28	26	44	28	33	32	28	36	36	27	26
CSR	99	263	101	100	74	100		527	1305	83	39	80	14	77	68	73	115
MSE	24510	24516	24520	24523	24516	24562		24670	24512	24509	24511	24507	24523	24509	24570	24515	24512
RMSE	31	33	47	35		88	55	88	43	42	51	47	48	52	46	39	35
Err3	32	37	58	43	100	65		100	49	48	65	55	58	58.7	46.6	45	22.5
SNR	18	17	14	14	12	6.1	16	6.2	14.2	15	13.2	13.4	13.7	11.9	15.2	14.7	15.4
Q	0.79	0.59	0.46	0.46	0.37	0.41	0.39	0.47	0.51	0.65	0.66	0.43	0.143	0.38	0.39	0.53	0.46
SSIN	0.81	0.51	0.33	0.33	0.43	0.29	0.21	0.29	0.28	0.55	0.60	0.36	0.41	0.41	0.47	0.57	0.42
SC	1.31	1.06	0.98	0.93	3.22	0.17		3.22	1.04	1.08	1.08	1.02	1.29	0.13	1.02	1.05	1.04
MD	125	127	188	169	241	241	199	241	171	157	255	180	249	203	175	149	129
NAE	0.22	0.18	0.23	0.22	0.49	0.49	1.01	0.49	0.24	0.17	0.26	0.24	0.26	0.31	0.19	0.17	0.16

selected based on the speckle index, C, and on the visual perception of the two experts. When C was minimally changing then the filtering process was stopped. As shown in Table 7.1, all filters reduced the C with the exception of the *DsFhomo* filter, which exhibited the worst performance as it moves the mean of the image, μ, to a darker gray level value, thus making the image darker. Filters that reduced the variance, σ^2, while preserving the mean, μ, and the median compared to the original image, were: *DsFlsmv*, *DsFwiener*, *DsFmedian*, *DsFls*, *DsFhomog*, *DsFhomo*, *DsFhmedian*, *DsFad*, *DsFncdif*, and *DsFwaveltc*. The contrast, of the image was increased by the filters *DsFgf4d* (enormously), *DsFhomo*, *DsFnlocal*, *DsFlsminsc*, *DsFls*, *DsFmedian*, *DsFhmedian*, and *DsFhomog* and it was decreased by the filters *DsFad*, *DsFwiener*, *DsFwaveltc*, and *DsFlsmv*. It is noted that filters *DsFgf4d*, *DsFlsmv*, and *DsFlsminsc* reduced C, increased CSR, *DsFlsmv* reduced the contrast, whereas *DsFlsminsc* increased the contrast.

The lower part of Table 7.1 tabulates the image quality evaluation metrics presented in Chapter 2 (Section 2.3), for the artificial carotid artery ultrasound image illustrated in Fig. 7.1. The quality metrics were calculated between the original (see Fig. 7.1a) and the despeckled images (see Fig. 7.1b–Fig. 7.1q). Best values were obtained for the *DsFlsmv*, *DsFlsminsc*, *DsFncdif*, *DsFwaveltc*, *DsFnlocal*, and *DsFhmedian* with higher SNR. Best values for the universal quality index, Q, and the structural similarity index, SSIN were obtained for the filters *DsFlsmv*, *DsFhmedian*, *DsFkuwahara*, and *DsFncdif*. The SC was best for the filters *DsFlsmv*, *DsFls*, and *DsFlsminsc*. The smallest MD values were given for the *DsFlsmv*, *DsFlsminsc*, and *DsFwaveltc*.

7.2   DESPECKLE FILTERING EVALUATION ON A PHANTOM IMAGE

Despeckle filtering was evaluated on a phantom carotid (that was introduced in Fig. 2.1a and used in introducing the filters in Chapters 3–6). Figure 7.3 shows the original phantom ultrasound images (see Fig. 7.3a) and the despeckled phantom images (see Fig. 7.3b–Fig. 7.3q) after the application of the despeckle filters *DsFlsmv*, *DsFwiener*, *DsFlsminsc*, *DsFmedian*, *DsFls*, *DsFhomog*, *DsFgf4d*, *DsFhomo*, *DsFhmedian*, *DsFKuwahara*, *DsFnlocal*, *DsFad*, *DsFsrad*, *DsFnldif*, *DsFncdif*, and *DsFwaveltc* for different window sizes (given in brackets) and number of iterations (shown in parentheses). Best results were obtained for the filters *DsFlsmv*, *DsFlsminsc*, *DsFwiener*, *DsFmedian*, *DsFhomog*, *DsFgf4d*, *DsFhomo*, *DsFhmedian*, *DsFkuwahara*, *DsFsrad*, and *DsFncdif*. The filters *DsFnlocal*, *DsFad*, *DsFsrad*, *DsFnldif*, and *DsFwaveltc* did not preserve the edges.

Moreover, it is shown in Fig. 7.3(i), Fig. 7.3(l), and Fig. 7.3(p), that the filters *DsFhomo*, *DsFnlocal*, and *DsFnldif* were noisy. Furthermore, diameter and area measurements were carried out for the anechoic cylinder of Fig. 7.3(a) and tabulated in Table 7.2. The following measurements were carried out: (i) two measurements of the diameter (where each measurement was perpendicular to the other), (ii) area of the anechoic cylinder using the average of the two diameter measurements, and (iii) the percentage area error difference between the original area and the area estimated in (ii).

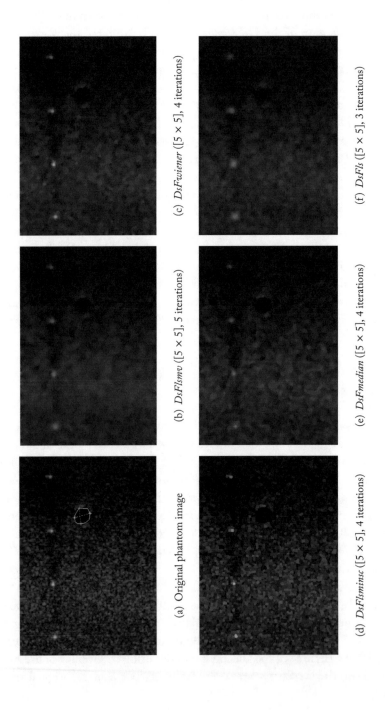

(a) Original phantom image

(b) *DsFlsmv* ([5 × 5], 5 iterations)

(c) *DsFrwiener* ([5 × 5], 4 iterations)

(d) *DsFlsminsc* ([5 × 5], 4 iterations)

(e) *DsFrmedian* ([5 × 5], 4 iterations)

(f) *DsFls* ([5 × 5], 3 iterations)

Figure 7.3: Original phantom image given in (a), and the application of 16 despeckle filters for different number of iterations and different pixel moving window sizes, shown in brackets, given in (b)–(q). See Table 7.3 for number of iterations and window size. (a) Original phantom image, (b) *DsFlsmv*, (c) *DsFrwiener*, (d) *DsFlsminsc*, (e) *DsFrmedian*, and (f) *DsFls*. (*Continues.*)

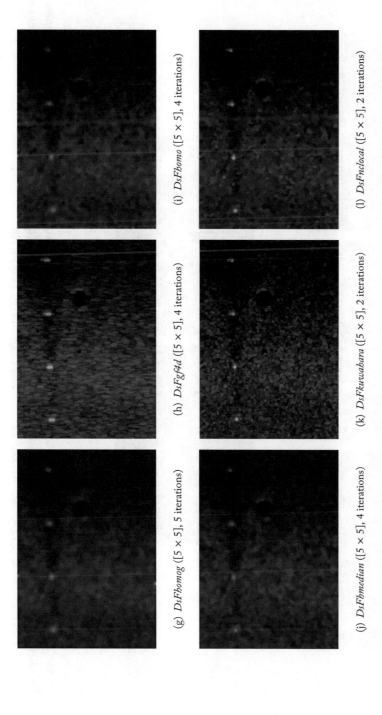

(g) *DsFhomog* ([5 × 5], 5 iterations)   (h) *DsFgf4d* ([5 × 5], 4 iterations)   (i) *DsFhomo* ([5 × 5], 4 iterations)

(j) *DsFbmedian* ([5 × 5], 4 iterations)   (k) *DsFkuvvahara* ([5 × 5], 2 iterations)   (l) *DsFnclocal* ([5 × 5], 2 iterations)

Figure 7.3: *(Continued.)* Original phantom image given in (a), and the application of 16 despeckle filters for different number of iterations and different pixel moving window sizes, shown in brackets, given in (b)–(q). See Table 7.3 for number of iterations and window size. (g) *DsFhomog*, (h) *DsFgf4d*, (i) *DsFhomo*, (j) *DsFbmedian*, (k) *DsFKuvvahara*, and (l) *DsFnlocal*. *(Continues.)*

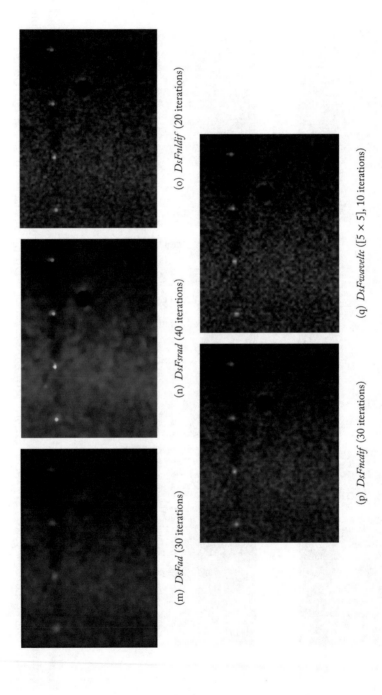

(m) *DsFad* (30 iterations)

(n) *DsFsrad* (40 iterations)

(o) *DsFnldif* (20 iterations)

(p) *DsFncdif* (30 iterations)

(q) *DsFwaveltc* ([5 × 5], 10 iterations)

Figure 7.3: *(Continued.)* Original phantom image given in (a), and the application of 16 despeckle filters for different number of iterations and different pixel moving window sizes, shown in brackets, given in (b)–(q). See Table 7.3 for number of iterations and window size. (m) *DsFad*, (n) *DsFsrad*, (o) *DsFnldif*, (p) *DsFncdif*, and (q) *DsFwaveltc.*

Table 7.2: Selected statistical features, image quality evaluation metrics and shape[1] measurements for Fig. 7.3 before and after despeckle filtering

Feature	original	Linear Filtering						Non-Linear Filtering					Diffusion				Wavelet
		DsF lsmv	DsF wiener	DsF lsminsc	DsF ls	DsF homog	DsF gf4d	DsF homo	DsF median	DsF hmedian	DsF Kuwahara	DsF nlocal	DsF ad	DsF srad	DsF nldif	DsF ncdif	DsF waveltc
No.of iterations		2	4	4	3	5	4	4	4	4	2	2	30	40	20	30	10
Window size	5x5	5x5	5x5	5x5	5x5	5x5	5x5	5x5	5x5	5x5	5x5	5x5	-	-	-	-	5x5
μ	36	35	36	36	36	37	47	34	35	36	36	37	35	50	36	36	36
Median	37	40	39	37	39	39	50	38	39	38	37	39	39	52	39	37	38
σ^2	21	17	18	22	16	17	22	19	18	18	21	23	17	27	19	20	20
σ^3	0.2	0.3	0.06	0.2	0.1	0.2	0.02	0.3	0.3	-0.07	0.16	2.6	0.3	0.2	0.1	1.3	0.1
σ^4	2.8	2	3	3	2.3	2.2	3	2	2	2.4	2.7	20.1	2	3	3.1	13.6	2.8
Contrast	76	4	6	114	5	3	50	13	7	11	79	112	3	57	19	71	29
C	58	48	50	51	44	46	47	56	51	50	58	58	49	54	53	56	55
CSR*100		0.5	0.02	0.09	0.1	0.4	4.7	0.95	0.4	0.08	0.05	0.08	0.4	5.8	0.01	0.1	0.01
MSE		1741	1754	1734	1765	1760	1776	1793	1751	1738	1744	1756	1765	1764	1732	1761	1739
RMSE		8.4	9.5	9.1	10.7	8.7	12.6	12.8	12.3	7.9	17.4	15.9	7.5	19.1	10.2	10.3	7.9
Err3		10.4	10.9	14.2	12.5	10.6	16.2	130	19.3	9.5	35.3	30.5	9.1	21.9	18.6	17.8	9.5
SNR		14.8	15.8	13.7	14.8	16.5	14.3	2.5	13.5	17.3	10.9	11.7	17.9	11.3	15.2	15.1	17.34
Q		0.73	0.36	0.22	0.27	0.52	0.63	0.29	0.37	0.65	0.56	0.46	0.70	0.19	0.70	0.63	0.66
SSIN		0.74	0.58	0.43	0.51	0.67	0.71	0.31	0.57	0.74	0.64	0.65	0.78	0.45	0.77	0.73	0.72
SC		1.31	1.11	1.22	1.27	1.08	0.71	0.06	1.08	1.09	0.88	0.95	1.08	0.57	1.02	1.05	1.04
MD		74	34	81	69	146	101	178	134	54	254	210	58	83	135	135	50
NAE		0.92	0.94	0.87	0.24	0.77	1.06	6.05	0.23	0.17	0.24	0.22	0.16	0.43	0.17	0.18	0.17
Diameter 1[1] [mm]	5.77	5.78	5.74	5.86	5.74	5.67	5.63	5.88	5.88	5.84	5.76	5.85	5.94	5.91	5.82	5.82	5.73
Diameter 2[1] [mm]	5.61	5.64	5.73	5.77	5.69	5.59	5.54	5.74	5.86	5.79	5.64	5.68	5.86	5.83	5.66	5.69	5.58
Area[1] [mm²]	17.87	17.93	18.01	18.26	17.95	17.68	17.54	18.24	18.43	18.26	17.90	18.10	18.53	18.43	18.02	18.07	17.76
%Area Error[1]		0.35	0.79	2.19	0.44	1.05	1.85	2.11	3.16	2.19	0.18	1.32	3.69	3.16	0.88	1.14	0.61

[1]Shape measurements refer to the anechoic cylinder of Fig. 7.3 (marked in Fig. 7.3a).

The upper part of Table 7.2 tabulates the statistical features, μ, median, σ^2, σ^3, σ^4, the contrast, speckle index, C (2.9), and contrast-speckle-radio, CSR (2.10), for the phantom image and the 16 despeckle filters illustrated in Fig. 7.3. As shown in Table 7.2, all filters, reduced C with the exception of the *DsFkuwahara* and *DsFnlocal* filters. The *CSR* is better for the *DsFsrad*, *DsFhomo*, *DsFlsmv*, *DsFhomog*, and *DsFad*. Filters that reduced the variance, σ^2, while preserving the mean, μ, and the median compared to the original image, were: *DsFhomo*, *DsFls*, *DsFwiener*, *DsFwaveltc*, *DsFad*, *DsFhomog*, *DsFmedian*, *DsFhmedian*, *DsFncdif*, and *DsFlsmv*. The contrast, of the image is increased by the filter *DsFlsminsc* and *DsFnlocal* (enormously), and preserved by *DsFncdif*, *DsFgf4d*, *DsFkuwahara*, *DsFsrad*, *DsFwaveltc*, and *DsFnldif*. It is decreased by the filters *DsFlsmv*, *DsFmedian*, *DsFwiener*, *DsFls*, *DsFhomog* and *DsFad*. It is noted that filters *DsFgf4d*, *DsFlsmv* and *DsFlsminsc* reduced C, *DsFgf4d* increased *CSR*, *DsFlsmv* reduced the contrast, whereas *DsFlsminsc* increased the contrast. The despeckled images of Fig. 7.3 were also assessed by the two experts. Filters that showed an improved smoothing after filtering, as assessed visually by the two experts, using visual perception criteria, are presented in the following order: *DsFwaveltc*, *DsFlsmv*, *DsFnldif*, *DsFsrad*, *DsFad*, *DsFncdif*, *DsFhmedian*, *DsFgf4d*, and *DsFmedian*. Filters that showed a blurring effect especially on the edges were: *DsFls*, *DsFlsminsc*, *DsFhomog*, *DsFhomo*, and *DsFwiener*.

The lower part of Table 7.2 tabulates the image quality evaluation metrics presented in Section 2.3, for the phantom carotid artery ultrasound image illustrated in Fig. 7.3. The image quality metrics were calculated between the original (see Fig. 7.3a) and the despeckled images (see Fig. 7.3b–Fig. 7.3q). Best values were obtained for the *DsFlsmv*, *DsFhmedian*, *DsFlsminsc*, *DsFnldif*, and *DsFwaveltc* with lower *RMSE* and *Err3*, and higher *SNR*. Best values for the universal quality index, Q, and the structural similarity index, *SSIN* were obtained for the filters *DsFlsmv*, *DsFhmedian*, *DsFad*, and *DsFnldif*. The *SC* was best for the filters *DsFlsmv*, *DsFls*, and *DsFlsminsc*. The smallest *MD* was given by *DsFwiener*, *DsFwaveltc*, and *DsFhmedian* filters, while for the *NAE* best values were obtained for the filters *DsFad*, *DsFwaveltc* and *DsFhmedian*. The filter *DsFhomo* increases *MSE*, *RMSE*, *Err3*, and *MD* enormously. Finally, the lowest part of Table 7.2 illustrates the manual measurements of diameter 1 and diameter 2 and the area of the anechoic cylinder in Fig. 7.3. The percentage area error for most of the filters was small of the order of a few percent.

7.3 DESPECKLE FILTERING EVALUATION ON REAL ULTRASOUND IMAGES AND VIDEO

Figure 7.4 shows an ultrasound image of the carotid artery together with the despeckled images. Table 7.3 tabulates the statistical features and image quality evaluation metrics of the despeckled images. The best visual results as assessed by the two experts were obtained for the filters *DsFlsmv*, *DsFlsminsc*, and *DsFkuwahara* whereas the filters *DsFgf4d*, *DsFad*, *DsFncdif*, and *DsFnldif* also showed good visual results but smoothed the image, loosing subtle details and affecting the edges. Filters that showed a blurring effect are the *DsFmedian*, *DsFwiener*, *DsFhomog*, and *DsFwaveltc*.

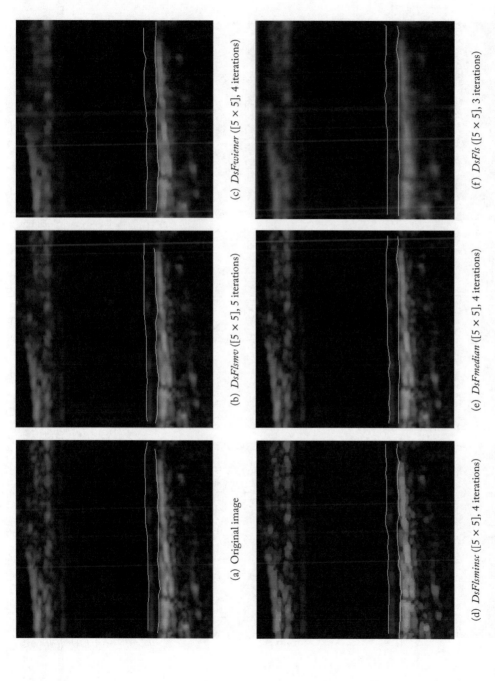

(a) Original image

(b) *DsFlsmv* ([5 × 5], 5 iterations)

(c) *DsFwiener* ([5 × 5], 4 iterations)

(d) *DsFlsminsc* ([5 × 5], 4 iterations)

(e) *DsFmedian* ([5 × 5], 4 iterations)

(f) *DsFls* ([5 × 5], 3 iterations)

Figure 7.4: Original ultrasound image of the carotid artery (2–3 cm proximal to bifurcation) given in (a), and the despeckled filtered images given in (b)–(q). The corresponding number of iterations and window size for each filter is given in Table 7.3. Also, the IMC segmentations derived using snakes as documented in [25] is demonstrated in the far wall of the image. In addition, the corresponding measurements of IMT are tabulated in the last row of Table 7.3. *(Continues.)*

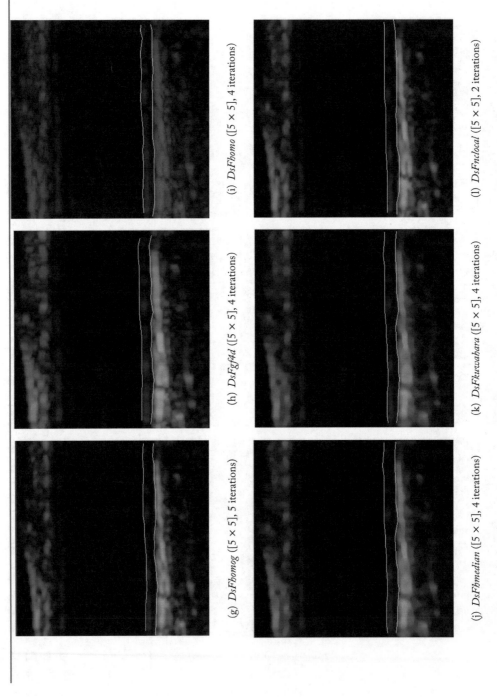

(g) *DsFhomog* ([5 × 5], 5 iterations)

(h) *DsFgf4d* ([5 × 5], 4 iterations)

(i) *DsFhomo* ([5 × 5], 4 iterations)

(j) *DsFbmedian* ([5 × 5], 4 iterations)

(k) *DsFkuwahara* ([5 × 5], 4 iterations)

(l) *DsFnclocal* ([5 × 5], 2 iterations)

Figure 7.4: *(Continued.)* Original ultrasound image of the carotid artery (2–3 cm proximal to bifurcation) given in (a), and the despeckled filtered images given in (b)–(q). The corresponding number of iterations and window size for each filter is given in Table 7.3. Also, the IMC segmentations derived using snakes as documented in [25] is demonstrated in the far wall of the image. In addition, the corresponding measurements of IMT are tabulated in the last row of Table 7.3. *(Continues.)*

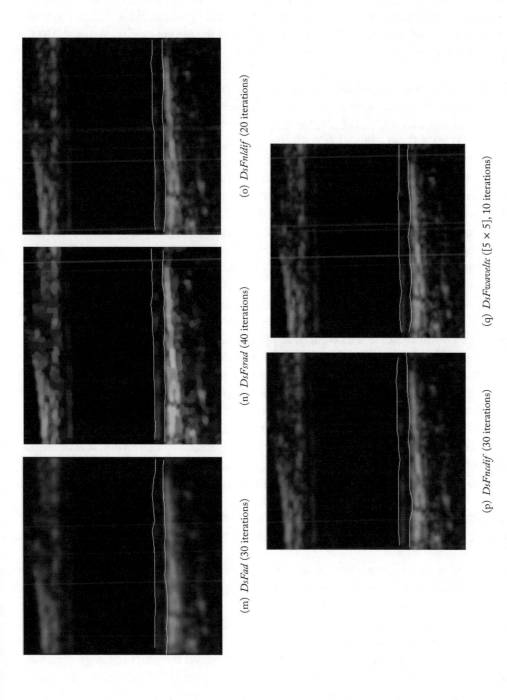

(m) *DsFad* (30 iterations)

(n) *DsFsrad* (40 iterations)

(o) *DsFnldif* (20 iterations)

(p) *DsFncdif* (30 iterations)

(q) *DsFwaveltc* ([5 × 5], 10 iterations)

Figure 7.4: *(Continued.)* Original ultrasound image of the carotid artery (2–3 cm proximal to bifurcation) given in (a), and the despeckled filtered images given in (b)–(q). The corresponding number of iterations and window size for each filter is given in Table 7.3. Also, the IMC segmentations derived using snakes as documented in [25] is demonstrated in the far wall of the image. In addition, the corresponding measurements of IMT are tabulated in the last row of Table 7.3.

Filters *DsFwiener, DsFls, DsFhomog*, and *DsFwaveltc* showed poorer visual results. The corresponding number of iterations and window size, for each filter is given in Table 7.3. The intima-media complex (IMC) was automatically segmented [25] and the intima-media thickness (IMT) measurements for each despeckled image are shown in millimeters in the last two rows of Table 7.3.

Additionally, the percentage difference (%) between the original and despeckled IMT measurements are shown. The smallest % difference was obtained by the filter *DsFlsminsc*.

Figure 7.5 shows two original ultrasound cardiac images in (a) and (g) and the despeckled images in (b), (c), (d), (d) (e), (f) and (g), (h), (i), (j), (k), (l) with filters, *DsFlsmv, DsFgf4d, DsFhmedian, DsFsrad*, and *DsFcdif*, respectively. The corresponding number of iterations and the moving sliding windows applied were the same as in Fig. 7.4. The best visual results as assessed by the two experts were obtained by the filters *DsFlsmv* (see Fig. 7.5b and Fig. 7.5h) and the filter *DsFhmedian* (see Fig. 7.5d and Fig. 7.5i). The rest of the filters showed also good visual results but smoothed the image loosing subtle details, affecting also the edges. Selected statistical and image quality metrics for Fig. 7.5 are presented in Table 7.4. Almost all filters preserved the mean, while the best *SSIN* was obtained for the filters *DsFhmedian* and *DsFlsmv*. Best values for *SNR*, and *NAE* were obtained by the *DsFhmedian* filter.

Figure 7.6 shows the original (see Fig. 7.6 (a), (c), and (e)) and despeckled (see Fig. 7.6 (b), (d), and (f)) frames numbered 1, 100, and 200 from an ultrasound CCA video consisting of 300 frames, with a width of 211 and a height of 256 pixels, captured at 30 frames/s. The *DsFlsmv* despeckle filter was iteratively applied for 2 iterations at consecutive video frames with a moving sliding window size of 5×5 pixels. The filtering was only applied to the luminance channel of the video (Y-channel). The speckle index (C according to (2.9)), for the original and despeckled frames was also calculated and it is given in parentheses. It is clear that C is reduced after despeckle filtering for all frames. Table 7.5 tabulates selected statistical features and image quality metrics for the original and the despeckled video frames for the despeckle filter *DsFlsmv*. It is observed that the effect of despeckle filtering is similar with that on images and also that there is variability in the tabulated statistical and quality metrics among the video frames investigated.

7.4 SUMMARY FINDINGS ON DESPECKLE FILTERING EVALUATION

Despeckle filtering is an important operation in the enhancement of ultrasound images of the carotid artery. In this book a total of 16 despeckle filters were presented and evaluated on artificial, phantom and real ultrasound images and videos.

As given in Table 7.6, filters *DsFlsmv* and *DsFhmedian* improved the statistical and texture features analysis, the measurements and shape features, the image quality evaluation and the optical perception evaluation. This was observed for both filters on the artificial image phantom and the real ultrasound images. The filter *DsFlsminsc* improved the statistical and texture image analysis and the image quality evaluation in real carotid ultrasound images. The filter *DsFgf4d*

Table 7.3: Selected statistical features, image quality evaluation metrics, and intima–media thickness (IMT) (mean $\pm$ std)[mm] measurements for images of Fig. 7.4 before and after despeckle filtering

Feature	original	Linear Filtering						Non-Linear Filtering					Diffusion				Wavelet
		DsF lsmv	DsF wiener	DsF lsminsc	DsF ls	DsF homog	DsF gf4d	DsF homo	DsF median	DsF hmedian	DsF Kuwahara	DsF nlocal	DsF ad	DsF srad	DsF nldif	DsF ncdif	DsF waveltc
No.of Iterations		2	4	4	3	5	4	4	4	4	2	2	30	50	20	20	10
Window size	5x5	5x5	5x5	5x5	5x5	5x5	5x5	5x5	5x5	5x5	5x5	5x5	-	-	-	-	5x5
μ	24	27	26	26	27	25	30	28	25	25	25	25	26	29	25	25	25
Median	21	19	21	9	14	12	13	9	9	9	6	8	10	9	8	9	8
σ^2	34	40	37	32	33	33	40	27	35	37	37	35	37	40	32	36	36
σ^3	1.99	2.42	2.41	1.79	2.11	2.28	1.91	2.0	2.18	2.38	2.18	2.22	2.47	1.86	2.19	1.99	2.28
σ^4	7.55	9.75	10.12	5.81	9.13	10.0	7.04	12	8.73	9.91	8.37	8.91	10.64	6.23	9.06	7.39	9.19
Contrast	143	392	321	49	186	96	279	106	175	212	284	174	296	111	59	126	318
C	1.45	1.44	1.42	1.23	1.24	1.30	1.32	0.96	1.42	1.48	1.48	1.48	1.42	1.38	1.28	1.44	1.44
CSR*100	-	1.47	1.36	0.87	1.53	0.57	3.02	2.07	0.49	0.69	0.69	0.51	0.49	2.42	0.52	0.50	0.50
MSE	-	1718	1717	1717	1719	1717	1719	1718	1716	1716	1717	1716	1718	1720	1717	1716	1717
RMSE	-	27	22	23	15	21	33	22	13	21	22	11	23	16	9	12	20
Err3	-	48	43	43	26	42	52	42	22	41	42	22	44	25	13	22	37
SNR	-	8	9	9	12	10	6	9	15	10	9	15	9	13	17	14	10
Q	-	0.83	0.67	0.69	0.66	0.63	0.68	0.52	0.81	0.66	0.71	0.78	0.61	0.58	0.59	0.78	0.71
SSIN	-	0.91	0.85	0.82	0.80	0.75	0.81	0.86	0.90	0.89	0.87	0.90	0.83	0.84	0.78	0.91	0.87
SC	-	0.75	0.86	0.85	0.99	0.84	0.69	0.94	0.95	0.87	0.93	0.96	0.65	0.64	1.1	1.03	0.88
MD	-	199	204	203	162	254	201	253	143	201	210	143	203	147	63	144	202
NAE	-	0.36	0.32	0.32	0.29	0.30	0.59	0.45	0.17	0.27	0.28	0.19	0.31	0.32	0.22	0.19	0.31
IMT$_{mean}$ [mm]	0.72	0.69	0.71	0.72	0.69	0.73	0.69	0.71	0.71	0.69	0.69	0.70	0.67	0.68	0.69	0.69	0.68
IMT$_{std}$ [mm]	0.90	0.11	0.9	0.12	0.8	0.9	0.7	0.7	0.7	0.8	0.8	0.70	0.7	0.8	0.9	0.8	0.8
Difference [%]	4.17	4.17	1.37	0.001	4.16	1.39	8.33	1.39	1.39	4.17	4.17	2.78	6.94	5.56	4.17	4.17	5.56

Difference [%]: Percentage difference [(Original_IMT$_{mean}$ − Despeckled_IMT$_{mean}$)/ Original_IMT$_{mean}$]*100% .

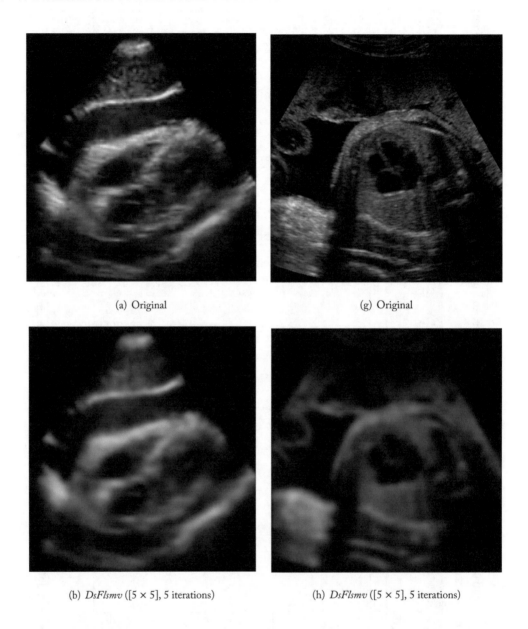

(a) Original   (g) Original

(b) *DsFlsmv* ([5 × 5], 5 iterations)   (h) *DsFlsmv* ([5 × 5], 5 iterations)

Figure 7.5: (a) and (e) Original cardiac ultrasound images (see also Chapter 1, Fig. 1.2) and the despeckled images with the filters *DsFlsmv, DsFgf4d, DsFhmedian, DsFsrad,* and *DsFcdif* given in the left and right columns for the images in (a) and (g), respectively. The corresponding number of iterations and the window size used were the same used in Fig. 7.4. *(Continues.)*

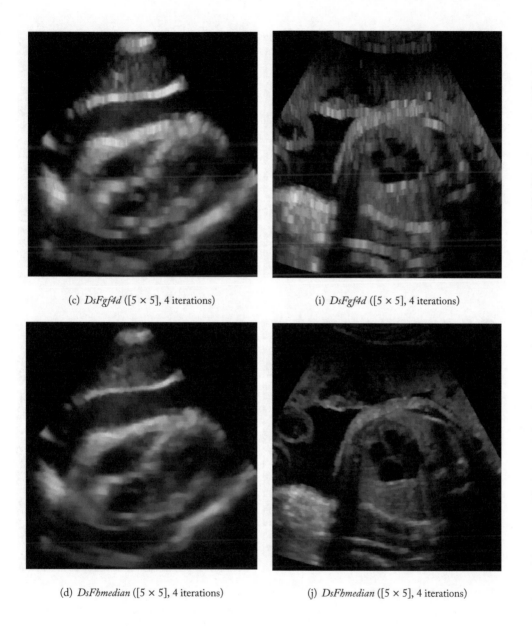

(c) *DsFgf4d* ([5 × 5], 4 iterations)

(i) *DsFgf4d* ([5 × 5], 4 iterations)

(d) *DsFhmedian* ([5 × 5], 4 iterations)

(j) *DsFhmedian* ([5 × 5], 4 iterations)

Figure 7.5: *(Continued.)* (a) and (e) Original cardiac ultrasound images (see also Chapter 1, Fig. 1.2) and the despeckled images with the filters *DsFlsmv, DsFgf4d, DsFhmedian, DsFsrad,* and *DsFcdif* given in the left and right columns for the images in (a) and (g), respectively. The corresponding number of iterations and the window size used were the same used in Fig. 7.4. *(Continues.)*

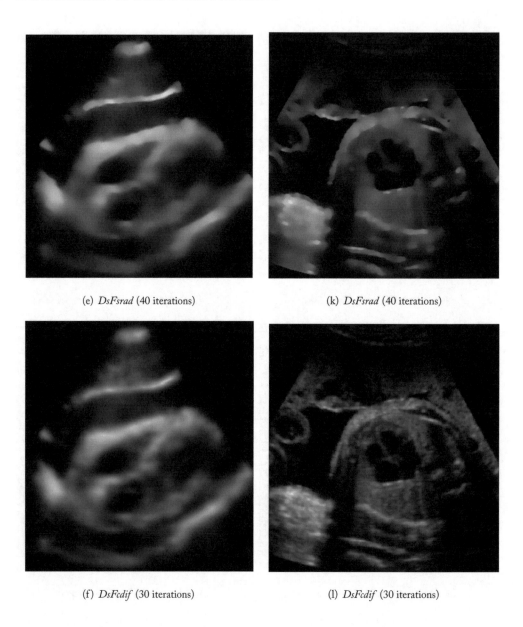

(e) *DsFsrad* (40 iterations)   (k) *DsFsrad* (40 iterations)

(f) *DsFcdif* (30 iterations)   (l) *DsFcdif* (30 iterations)

Figure 7.5: *(Continued.)* (a) and (e) Original cardiac ultrasound images (see also Chapter 1, Fig. 1.2) and the despeckled images with the filters *DsFlsmv, DsFgf4d, DsFhmedian, DsFsrad*, and *DsFcdif* given in the left and right columns for the images in (a) and (g), respectively. The corresponding number of iterations and the window size used were the same used in Fig. 7.4.

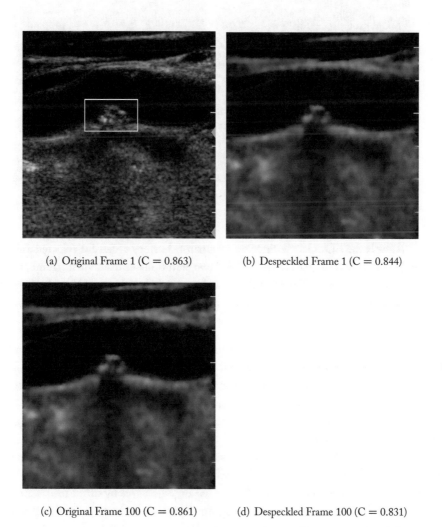

(a) Original Frame 1 (C = 0.863)   (b) Despeckled Frame 1 (C = 0.844)

(c) Original Frame 100 (C = 0.861)   (d) Despeckled Frame 100 (C = 0.831)

Figure 7.6: Despeckle filtering of a carotid artery video for selected frames. The despeckle filter *Ds-Flsmv* was iteratively applied for two iterations at each video frame, using a sliding moving window of size [5 × 5]. The carotid plaque is indicated with an ROI in the first frame at the far wall of the artery. The despeckle index (*C*) is also given for the corresponding despeckled frames. *(Continues.)*

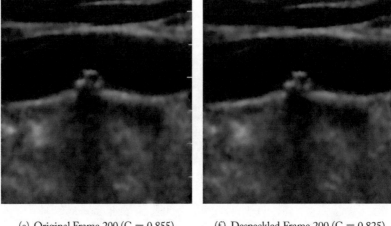

(e) Original Frame 200 (C = 0.855)   (f) Despeckled Frame 200 (C = 0.825)

Figure 7.6: *(Continued.)* Despeckle filtering of a carotid artery video for selected frames. The despeckle filter *DsFlsmv* was iteratively applied for two iterations at each video frame, using a sliding moving window of size [5 × 5]. The carotid plaque is indicated with an ROI in the first frame at the far wall of the artery. The despeckle index (*C*) is also given for the corresponding despeckled frames.

is suitable for improving the visual evolution in real ultrasound images. The *DsFhomo* showed a small marginal improvement for the statistical image features. Filters *DsFnldif* and *DsFwaveltc* showed similar performance by improving the measurements and shape features and the image quality evaluation of real ultrasound carotid images.

The *DsFlsmv* filter, which is a simple filter, is based on local image statistics gave very good performance. It was first introduced in [36, 38, 48] by Jong-Sen Lee and coworkers and was tested on a few SAR images with very satisfactory results. It was also used for SAR imaging in [47] and image restoration in [50], again with very satisfactory results.

Phantom images were used in this book (see Fig. 7.2) and by other researchers to evaluate despeckle filtering in carotid ultrasound imaging. Specifically in [3] a synthetic carotid ultrasound image of the CCA was used to evaluate the *DsFsrad* filtering (speckle reducing anisotropic filtering) which was compared with the *DsFlsmv* (Lee filter) [38] and the *DsFad* filter (conventional anisotropic diffusion) [39]. The edges of the phantom image used in [3], were studied and it was shown that the *DsFsrad* does not blurred edges as with the other two despeckle filtering techniques evaluated (*DsFlsmv* and *DsFad*).

Despeckle filtering was investigated by other researchers and also in our study, on an artificial carotid image (Fig. 7.1), [7, 29] on line profiles (Fig. 7.2) of different ultrasound images, [2, 3, 7, 28, 60], on phantom ultrasound images [3, 5, 40], SAR images [51, 54–56], real longitudinal ultrasound images of the carotid artery (Fig. 7.4) [3, 7, 29], and cardiac ultrasound

Table 7.4: Selected statistical features and image quality evaluation metrics for images of Fig. 7.5a (-/) and Fig. 7.5g (/-), before and after despeckle filtering

Feature	original	*DsFlsmv*	*DsFgf4d*	*DsFhmedian*	*DsFsrad*	*DsFncdif*
No.of Iterations		2	4	4	50	20
Window size		5x5	5x5	5x5	-	-
μ	42/ 37	43 / 33	53 / 49	44 / 38	44 / 37	41 / 37
Median	27/ 18	29 / 22	60 / 42	27 / 29	32 / 31	30 / 31
σ^2	48 / 33	41 / 33	51 / 45	41 / 34	45 / 35	43 / 33
σ^3	1.18 / 1.14	1.03 / 1.04	0.99 / 0.71	1.14 / 1.34	1.07 / 1.1	1.08 / 1.25
σ^4	3.87 3.73	3.34 / 3.58	3.28 / 2.76	3.65 / 4.88	3.52 / 3.59	3.58 / 4.42
Contrast	91 / 82	87 / 56	75 / 54	81 / 98	85 / 37	89 / 95
C	114 / 112	95 / 99	96 / 92	93 / 89	98 / 95	95 / 89
CSR*100	-	0.51 / 0.01	0.32 / 0.44	0.034 / 0.21	0.45/ 0.47	0.29 / 0.36
MSE	-	3612 / 2438	3618 / 2436	3611 / 2435	3613/ 2442	3616 / 2441
RMSE	-	6.2 / 12.2	7.1 / 28	6.76 / 15	6.27 / 13	6.89 / 11.5
Err3	-	54 / 17	48 / 38	10 / 32	54 / 17	19 / 26
SNR	-	11 / 15	11 / 10	22 / 14	11 / 16	18 / 16
Q	-	0.78 / 0.67	0.61 / 0.59	0.69 / 0.59	0.58 / 0.57	0.60 / 0.61
SSIN	-	0.89 / 0.71	0.74 / 0.72	0.91 / 0.79	0.82 / 0.75	0.85 / 0.85
SC	-	0.93 / 1.12	0.69 / 0.55	0.98 / 0.96	0.84 / 0.02	1.1 / 1.01
MD	-	225 / 217	221 / 179	79 / 208	245 / 142	102 / 187
NAE	-	0.19 / 0.25	0.28 / 0.53	0.09 / 0.19	0.21 / 0.25	0.14 / 0.15

images (see Fig.7.5). There are only two studies [29], and [7] where despeckle filtering was investigated on real, and artificial longitudinal ultrasound image of the carotid artery. Four different despeckle filters were applied in [29], namely the *DsFlsmv* [38], Frost [47], *DsFad* [27], and a *DsFsrad* filter [29]. The despeckle window used for the *DsFlsmv*, and Frost filters was 5 × 5 pixels. To evaluate the performance of these filters, the mean and the standard deviation were used, which were calculated in different regions of the carotid artery image, namely in lumen, tissue, and at the vascular wall. The mean gray level values of the original image for the lumen, tissue and wall regions were 1.03, 5.31, and 22.8, whereas the variance were 0.56, 2.69, and 10.61. The mean after despeckle filtering with the *DsFsrad* gave brighter values for the lumen and tissue. Specifically the mean for the lumen, tissue, and wall for the *DsFsrad* was (1.19, 6.17, 18.9), *DsFlsmv* (1.11, 5.72, 21.75), Frost (1.12, 5.74, 21.83), and *DsFad* (0.90, 4.64, 14.64). The standard deviation for the *DsFsrad* gave lower values (0.15, 0.7, 2.86) when compared with Lee (0.33, 1.42, 5.37), Frost (0.32, 1.40, 5.30), and *DsFad* (0.20, 1.09, 3.52). It was thus shown that the *DsFsrad* filter preserves the mean and reduces the variance. The number of images investigated

Table 7.5: Selected statistical features and image quality evaluation metrics for images of Fig. 7.6 for the despeckle filter *DsFlsmv* on selected video frames before and after despeckle filtering

Feature	original	*DsFlsmv*	original	*DsFlsmv*	original	*DsFlsmv*
No.of Iterations	Frame 1		Frame 100		Frame 200	
μ	40.14	44.88	44.64	44.79	44.86	45.06
Median	38.23	45.35	42.05	45.05	42.34	45.81
σ^2	34.42	37.43	38.99	37.34	38.86	37.22
σ^3	0.65	0.72	0.61	0.72	0.62	0.73
σ^4	2.99	4.01	2.69	4.03	2.72	4.09
Contrast	215	307	231	305	234	309
C	85	82	87	84	87	83
CSR*100	-	1.1	-	0.36	-	0.65
MSE	-	3281	-	3780	-	3787
RMSE	-	4.76	-	22.4	-	11.4
Err3	-	9.58	-	41.77	-	13.51
SNR	-	25	-	12	-	18
Q	-	0.99	-	0.65	-	0.71
SSIN	-	0.99	-	0.77	-	0.84
SC	-	1.05	-	1.11	-	1.18
MD	-	51	-	254	-	82
NAE	-	0.002	-	0.22	-	0.15

in [29], was very small, visual perception evaluation by experts was not carried out, as well as only two statistical measures were used to quantitatively evaluate despeckle filtering, namely the mean, and the variance before and after despeckle filtering as explained above. We believe that the mean and the variance used in [29] are not indicative and may not give a complete and accurate evaluation result as in [7]. Furthermore, despeckle filtering was investigated by other researchers on ultrasound images of, heart [3], pig heart [28], pig muscle [108], kidney [5], liver [53], echocardiograms [52], CT lung scans [85], MRI images of brain [109], brain X-ray images [110], SAR images [51], and real-world images [57].

Line plots, as used in our study (see Fig. 7.2), were also used in few other studies to quantify despeckle filtering performance. Specifically in [110], a line profile through the original and the despeckled ultrasound image of kidney was plotted, using adaptive Gaussian filtering. In [1] line profiles were plotted on four simulated and 15 ultrasound cardiac images of the left ventricle, in order to evaluate the *DsFmedian* filter. In another study [3], line profiles through one phantom, one heart, one kidney, and one liver ultrasound image, were plotted where an adaptive shrinkage weighted median [53, 57], *DsFwaveltc* (wavelet shrinkage) [71], and wavelet shrinkage

Table 7.6: Summary findings of despeckle filtering in an artificial carotid image (A), a phantom image (P), and real ultrasound image (C)

Despeckle Filter	Statistical and Texture Features	Measurements and shape features	Image Quality Evaluation	Optical Perception Evaluation
	A/P/C	A/P/C	A/P/C	A/P/C
Linear Filtering				
DsFlsmv	✓/✓/✓	-/✓/✓	✓/✓/✓	-/✓/✓
DsFlsminsc	-/-/✓	-/-/-	-/-/✓	-/-/-
Non-Linear Filtering				
DsFgf4d	-/-/-	-/-/-	-/-/-	✓/-/✓
DsFhomo	✓/-/-	-/-/-	-/-/-	-/-/-
DsFhmedian	✓/-/✓	✓/-/-	✓/✓/✓	✓/-/-
Diffusion Filtering				
DsFnldif	-/-/-	-/-/✓	-/-/✓	-/-/-
Wavelet Filtering				
DsFwaveltc		-/-/✓	-/-/✓	-/-/-

coherence enhancing [52] models were used and compared with a nonlinear coherent diffusion model [71]. Finally, in [28], line plots were used in one artificial computer simulated image, and one ultrasound image of pig heart, where an adaptive shrinkage weighted median filter [53, 57], a multiscale nonlinear thresholding without adaptive filter pre-processing [28], a wavelet shrinkage filtering method [71], and a proposed adaptive nonlinear thresholding with adaptive pre-processing method [28], were evaluated. In all of the above studies, visual perception evaluation by experts, statistical and texture analysis, on multiple images, as performed in our study, was not performed.

To the best of our knowledge, an exhaustive review on despeckle filtering in ultrasound images has not yet been carried out. However, different aspects of despeckle filtering surveyed in [111, 112] where various speckle characterization methods were introduced. It was also shown in [111] that the speckle pattern from one specific area will ideally be unique and, therefore, it is identifiable. This makes speckle regarded as a fundamental source of information which is to be exploited for driving different applications. These applications include adaptive speckle suppression or filtering as shown in this book, speckle decorrelation [114], classification [115], free-hand 3D reconstruction of ultrasound images [116], elastography [117], speckle tracking echocardiography [118], tissue characterization [112], ultrasound image deconvolution and segmentation [119, 120], and other methods where data-driven terms are crucial, e.g., level-sets [121], graph-cuts [122] or power watershed [36]. Once the speckle regions of interest are identified, it is also possible to track their patterns in the case of a continuously moving organ, such as the heart. This procedure is habitually referred to as "speckle tracking" whose essence is to compare two consecutive image frames and to follow the speckle regions of interest.

Most of the papers published in the literature for video filtering are limited to the reduction of additive noise, mainly by frame averaging. More specifically, in [123] the Wiener filtering method was applied to 3D image sequences for filtering additive noise, but results have not been thoroughly discussed and compared with other methods. The method was superior when compared to the purely temporal operations implemented earlier [124]. The pyramid thresholding method was used in [124], and wavelet-based additive denoising was used in [125] for additive noise reduction in image sequences. In another study [126], the image quality and evaluation metrics, were used for evaluating the additive noise filtering and the transmission of image sequences through telemedicine channels. An improvement of almost all the quality metrics extracted from the original and processed images was demonstrated. An additive noise reduction algorithm, for image sequences, using variance characteristics of the noise was presented in [127]. Estimated noise power and sum of absolute difference employed in motion estimation were used to determine the temporal filter coefficients. A noise measurement scheme using the correlation between the noisy input and the noise-free image was applied for accurate estimation of the noise power. The experimental results showed that the proposed noise reduction method efficiently removes noise. An efficient method for movie denoising that does not require any motion estimation was presented in [128]. The method was based on the fact that averaging several realizations of a random variable reduces the variance. The method was unsupervized and was adapted to denoise image sequences with an additive white noise while preserving the visual details on the movie frames. Very little attention has been paid to the problem of missing data (impulsive distortion) removal in image sequences. In [129] a 3D median filter for removing impulsive noise from image sequences was developed. This filter was implemented without motion compensation and so the results did not capture the full potential of these structures. Further, the median operation, although quite successful in the additive noise filtering in images, invariably introduces distortion when filtering of image sequences [129]. This distortion primarily takes the form of blurring fine image details.

The application of despeckle filters (see Chapters 2–5), the extraction of texture features (see Section 2.3), the calculation of image quality metrics (see Section 2.3), and the visual perception evaluation by experts (see Section 2.3) may also be applied to video. The video is broken into frames, which can then be processed one by one and then grouped together to form the processed video. Preliminary results for the application of despeckle filtering in an ultrasound carotid and an ultrasound cardiac video were presented in Section 7.3. However, significant work still remains to be carried out.

CHAPTER 8

Summary and Future Directions

8.1 SUMMARY

Despeckle filtering has been a rapidly emerging research area in recent years. The basic principles, theoretical background, and algorithmic steps of a representative set of despeckle filters were covered in this book. Moreover, selected representative applications of image despeckling covering a variety of ultrasound image processing tasks will be presented in a companion monograph [131]. Most importantly, a despeckle filtering and evaluation protocol for imaging and video is documented in Table 8.1. The source code of the algorithms discussed in this book has been made available on the web, thus enabling researchers to more easily exploit the application of despeckle filtering in their problems under investigation.

A total of 16 different despeckle filters were documented in this book based on linear filtering, nonlinear filtering, diffusion filtering and wavelet filtering. We have evaluated despeckle filtering on ultrasound phantom, artificial, and real ultrasound images of the carotid artery bifurcation, based on visual evaluation by two medical experts, texture analysis measures, and image quality evaluation metrics. The results of this study, suggest that the first-order statistics despeckle filter *DsFlsmv*, may be applied on ultrasound images to improve the visual perception and automatic image analysis.

Furthermore, despeckle filtering was used as a pre-processing step for the automated segmentation of the IMT [25] and the carotid plaque [9], followed by the carotid plaque texture analysis, and classification. Despeckle filters *DsFlsmv*, *DsFlsminsc*, and *DsFgf4d* gave the best performance for the segmentation tasks [113]. It was shown in [25] that when normalization and speckle reduction filtering was applied on ultrasound images of the carotid artery prior to IMT segmentation, the automated segmentation measurements were closer to the manual measurements. Our findings showed promising results, however, further work is required to evaluate the performance of the suggested despeckle filters at a larger scale as well as their impact in clinical practice. In addition, the usefulness of the proposed despeckle filters, in portable ultrasound systems and in wireless telemedicine systems still has to be investigated. These topics will be covered in detail in a companion monograph [130].

For those readers whose principal need is to use existing image despeckle filtering technologies and apply them on different type of images, there is no simple answer regarding which specific filtering algorithm should be selected without a significant understanding of both the filtering fundamentals, and the application environment under investigation. A number of issues would need to be addressed. These include availability of the images to be processed/analyzed, the required level of filtering, the application scope (general-purpose or application-specific), the application goal (for extracting features from the image or for visual enhancement), the allowable computational complexity, the allowable implementation complexity, and the computational requirements (e.g., real-time or offline). We believe that a good understanding of the contents of this book can help the readers make the right choice of selecting the most appropriate filter for the application under development. Furthermore, the despeckle filtering evaluation protocol documented in Table 8.1 could also be exploited.

Table 8.1: Despeckle filtering and evaluation protocol.

1	*Recording of ultrasound images*: Ultrasound images are acquired by ultrasound equipment and stored for further image processing. Regions of interest (ROI's) could be selected for further processing.
2	*Normalize the image*: The stored images may be retrieved and a normalized procedure may be applied (as described for example in Section 2.2).
3	*Apply despeckle filtering*: Select the set of filters to apply despeckling together with their corresponding parameters (like moving window size, iterations, and other).
4	*Texture features analysis*: After despeckle filtering the user may select ROI's (i.e. the plaque or the area around the intima-media thickness) and extract texture features. Distance metrics between the original and the despeckled images may be computed (as well as between different classes of images if applicable).
5	*Compute image quality evaluation metrics*: On the selected ROI's compute image quality evaluation metrics between the original noisy and the despeckled images.
6	*Visual quality evaluation by experts*: The original and/or despeckled images may be visually evaluated by experts.
7	*Select the most appropriate despeckle filter/filters:* Based on steps 3 to 6 and construct a performance evaluation table (for example, see Table 7.6) and select the most appropriate filter(s) for the problem under investigation.

8.2   FUTURE DIRECTIONS

The despeckle filtering algorithms, and the measures for image quality evaluation introduced in this book can also be generalized and applied to other image and video processing applications. Only a small number of filtering algorithms and image quality evaluation metrics were investigated in this book, and numerous extensions and improvements can be envisaged.

In general, the development of despeckle filtering algorithms for image despeckling, is a well investigated field and many researchers have been involved in this subject, but there is still not an appropriate method proposed, which will satisfy both the visual and the automated interpretation of image processing and analysis tasks. Most importantly more comparative studies of despeckle filtering are necessary, where different filters could be evaluated by multiple experts as well as based on image quality and evaluation metrics as also proposed in this book.

Additionally, the issue of video despeckling is still in its infancy, although it is noted that the proposed methodology and filtering algorithms documented in this book may be also investigated in video sequences (by frame filtering). There are many issues related to video despeckle filtering that remain to be solved. In general, the development of a multiplicative model based on video sequences is required, since most of the models developed for video filtering were for additive noise [126–129].

Despeckle filtering may be also applied in the pre-processing of ultrasound images for other organs, including the detection of hyperechoic or hypoechoic lesions in the kidney, liver, spleen, thyroid, kidney, echocardiographic images, mammography and other. It may be particularly effective when combined with harmonic imaging, since both can increase tissue contrast. Speckle reduction can also be extremely valuable when attempting to fuse ultrasound with Computed Tomography (CT), MRI, Positron Emission Tomography (PET) or Optical Coherence Tomography (OCT) images. For example, when a lesion is suspected on a CT scan but it is not clearly visible ultrasound despeckle filtering can be applied in order to accentuate subtle borders that may be masked by speckle. It was also shown that speckle reduction filtering may enhance performance in image and video encoding and transmission, as it was shown in [130], where videos of atherosclerotic carotid plaques were transmitted over a wireless telemedicine channel. Previous surveys of encoding methods in other areas have been reported in [131–134].

Ultrasound imaging instrumentation, linked with imaging hardware and software technology have been rapidly advancing in the last two decades. Although these advanced imaging devices produce higher quality images and video, the need still exists for better image and video processing techniques including despeckle filtering. Towards this direction, it is anticipated that the effective use of despeckle filtering (by exploiting the filters and algorithms documented in this book) will greatly help in producing images with higher quality. These images that would be not only easier to visualise and to extract useful information, but would also enable the development of more robust image pre-processing and segmentation algorithms, minimizing routine manual image analysis and facilitating more accurate automated measurements of both industrially and clinically-relevant parameters.

It should be noted that the processing time of the proposed algorithmic methods presented in this book, could be further reduced by applying despeckle filtering only on selected areas of the image. Furthermore, software optimization methods (i.e., the MATLAB™ software optimization toolbox) could be investigated for increasing the performance of the proposed IDF and VDF software systems. Finally, it should be noted that the proposed methods could also be applied to other applications, such as echocardiography but a direct comparison of the results produces with this study will not be possible as different results will be produced with a different database.

In the companion monograph of this book [135], selected applications on despeckle filtering for images and videos will be presented where a more detailed evaluation of despeckle filtering based on texture features, image quality metrics and multiple observer evaluation on a large number of images and videos will be presented. More specifically, applications of despeckle filtering on ultrasound imaging and video of the intima-media complex, plaque segmentation and texture analysis will be presented. Moreover, the usefulness of despeckle filtering in reducing bandwidth needs in an ultrasound video telemedicine platform will be presented.

APPENDIX A

Appendices

Appendix Section A.1 contains a listing of all the functions included in the image despeckle filtering (IDF) toolbox, as introduced in Table 1.3 and Section 2.2 of this book. The IDF toolbox, which can be downloaded from `http://www.medinfo.cs.ucy.ac.cy/`, includes all the functions used for the texture analysis, (see also companion monograph), as well as for the image quality and evaluation presented in Section 2.3. Appendix A.2 lists all the functions included in the video despeckle filtering (VDF) toolbox for video analysis introduced in Section 2.2. Appendix A.3 presents an example in Matlab™ code for a complete application of despeckling, image quality evaluation and texture analysis. All page numbers listed refer to pages in the book, indicating where a function is first used and illustrated.

A.1 DESPECKLE FILTERING, TEXTURE ANALYSIS, AND IMAGE QUALITY EVALUATION TOOLBOX FUNCTIONS

The following MATLAB functions are grouped in categories as presented in Table 1.3 of this book.

Function Category and Name	Description	Page or Other Location
Linear Filtering		
DsFlsmv	Mean and variance local statistics despeckle filter	p. 43, Algorithm 3.1
DsFlsmvsk1d	Minimum variance homogenous 1D mask despeckle filter	p. 56
DsFlsmvsk2d	Minimum variance, higher moments local statistics despeckle filter	p. 58, Algorithm 3.3
DsFlsminsc	Minimum speckle index homogenous mask despeckle filter	p. 61, Algorithm 3.4
DsFwiener	Wiener despeckle filter	p. 56, Algorithm 3.2
Non-Linear Filtering		
DsFmedian	Median despeckle filter	p. 68, Algorithm 4.1

DsFls	Linear scaling of the gray level values despeckle filter	p. 68
DsFca	Linear scaling of the gray-levels despeckle filter	p. 69, Algorithm 4.2
DsFlecasort	Linear scaling and sorting despeckle filter	p. 71
DsFgf4d	Geometric despeckle filtering	p. 74, Algorithm 4.4
DsFhomog	Most homogeneous neighborhood despeckle filter	p. 71, Algorithm 4.3
DsFhomo	Homomorphic despeckle filtering	p. 74
DsFhmedian	Hybrid median despeckle filter	p. 80, Algorithm 4.6
DsFkuwahara	Kuwahara nonlinear despeckle filtering	p. 82, Algorithm 4.7
DsFnlocal	Non-local despeckle filtering	p. 84, Algorithm 4.8
Diffusion Filtering		
DsFad	Perona and Malik diffusion filter	p. 85
DsFsrad	Speckle reducing anisotropic diffusion filter	p. 87, Algorithm 5.1
DsFnldif	Nonlinear coherent diffusion despeckle filter	p. 93
DsFncdif	Nonlinear complex diffusion filtering	p. 95
Wavelet Filtering		
DsFwaveltc	Wavelet despeckle filtering	p. 99, Algorithm 6.1

The following image texture analysis MATLAB functions (also presented in the companion monograph), which are included in the IDF [13] and VDF [15] toolboxes, for image and video analysis are here below described:

Function Category and Name	Description	Page or Other Location
DsTnwfos	First-Order Statistics (FOS) (features 1–5)	Website
	Texture Analysis Functions	
DsTnwsgldm	Haralick Spatial Gray Level Dependence Matrices (SGLDM) (6–31)	Website
DsTnwgldmc	Gray Level Difference Statistics (GLDS) (32–35)	Website
DsTnwngtdmn	Neighbourhood Gray Tone Difference Matrix (NGTDM) (36–40)	Website
DsTnwsfm	Statistical Feature Matrix (SFM) (41–44)	Website
DsTnlaws	Laws Texture Energy Measures (TEM) (45–50)	Website
DsTfdta2	Fractal Dimension Texture Analysis (FDTA) (51–54)	Website
DsTfps	Fourier Power Spectrum (FPS) (55–56)	Website

DsTfshape2	Shape (x, y, area, perim, perim^2/area) (57–61)	Website
DsTintens2	Intensity difference vector with step s	Website
DsTleast	Estimation of the curve slope using least squares	Website
DsTresol2	Multiple resolution feature extraction	Website
DsTexfeat	Main texture analysis function	Website

Website: The Matlab™ code can be downloaded from `http://www.medinfo.cs.ucy.a` `c.cy` as well as from Researchgate.

The following image quality evaluation MATLAB functions are given as presented in the companion monograph to this book and are included in the IDF [13] and VDF [15] toolboxes for image and video analysis.

Function Category and Name	Description	Page or Other Location
Quality Evaluation		
DsQEgae	Geometric average error	p. 35
DsQEmse	Mean square error	p. 38
DsQEsnr	Signal-to-noise ratio	p. 36
DsQErmse	Randomized mean square error	p. 35
DsQEpsnr	Peak signal-to-noise ratio	p. 36
DsQEminkowski	Minkowski metrics, 3rd (M3) and 4th (M4) moments	p. 35
DsQEimg_qi	Universal quality index	p. 36
DsQEssim_index	Structural similarity index	p. 36
DsQEget_dir_files	Get directory files	Website
DsQE_quality_evaluation	Main quality evaluation program	Website
DsQmetrics	Function for running all above Quality Evaluation Metrics	Website

Website: The Matlab™ code can be downloaded from `http://www.medinfo.cs.ucy.a` `c.cy`

A.2   DESPECKLE FILTERING, TEXTURE ANALYSIS, AND VIDEO (VDF TOOLBOX) QUALITY EVALUATION TOOLBOX FUNCTIONS

The following Matlab™ functions are included in the VDF toolbox [15] used for video despeckle filtering.

Function Category and Name	Description	Page or Other Location
Linear filtering		
DsFlsmv	Mean and variance local statistics despeckle filter	p. 43, Algorithm 3.1
Nonlinear Filtering		
DsFhmedian	Hybrid median despeckle filter	p. 80, Algorithm 4.6
DsFkuwahara	Kuwahara nonlinear despeckle filtering	p. 82, Algorithm 4.7
Diffusion Filtering		
DsFsrad	Speckle reducing anisotropic diffusion filter	p. 88, Algorithm 5.1
Wavelet Filtering		
DsFwaveltc	Wavelet despeckle filtering	p. 99, Algorithm 6.1

A.3 EXAMPLES OF RUNNING THE DESPECKLE FILTERING TOOLBOX FUNCTIONS

Matlab™ code for the **DsQmetrics.m** function:

The following code sequence will read an image and apply the *DsFlsmv* despeckle filter on the image iteratively five times, by using a moving sliding window of 7×7 pixels. The texture features as well as the image quality metrics between the original and the despeckled images, are calculated with the code in Algorithm 3.1, and stored in the variable matrix A and B, respectively, (see also Chapter 3). The image quality metrics between the original and the despeckled images are stored in the matrix M.

Code A.1 Matlab™ code for the **DsQmetrics.m** function.

```
function f=DsQmetrics(I,K);
% I: Original input noisy image
% K: Despeckled input Image
% f: Matrix with image quality metrics

I=double(I); K=double(K);
gaer = DsQgae(I,K);
metrics= [gaer];

% calculate the mean square error mse
mser=DsQmse(I,K);
metrics=[metrics, mser];

% calculate the signal-to-noise radio snr
snrad=DsQsnr(I,K);
metrics=[metrics, snrad];

% calculate the square root of the mean square error
rmser=DsQrmse(I, K);
metrics=[metrics, rmser];

% Calculate the peak-signal-to-noise radio
psnrad=DsQpsnr(I,K);
metrics=[metrics, psnrad];

% Calculate the Minkwofski measure
[M3, M4] = DsQminkowski(I, K);
metrics=[metrics, M3, M4];

% Calculate the universal quality index
[quality, quality_map] = DsQimg_qi(I,K);
metrics=[metrics, quality];

% Calculate the structural similarity index
[mssim, ssim_map] = DsQssim_index(I, K);
metrics=[metrics, mssim];

% calculate aditional metrics
[MSE,PSNR,AD,SC,NK,MD,LMSE,NAE,PQS]= DsQiq_measures(I,K);
metrics=[metrics, AD, SC, NK, MD, LMSE, NAE, PQS];
 f= metrics;
```

Code A.2

```
% Read the image original.tif and store it in variable image
image = imread ('original.tif');

% Apply the despeckle filter DsFlsmv on the image using a sliding moving window of 7x7 pixel,
%iteratively 5 times
despeckle = DsFlsmv (image, [7 7], 5);
% Show the original and the despeckled images on the screen
figure, imshow (image); figure, imshow (despeckle);

% Calculate the texture features for the original and the despeckled images
Orig_textfeat = DsTexfeat (image);
Desp_textfeat = DsTexFeat (despeckle);

% Save the extracted features of the original and despeckled images in the mat files A and B
save Orig_textfeat A;
save Desp_textfeat B;

% Calculate the image quality evaluation metrics between the original and the despeckled images
% and save them in a matrix M
M = DsQmetrics (image, despeckle);

%The mat files A, B can then be loaded into the MATLAB workspace for opening, reading and
storing the features and image quality metrics. This can be made by double clicking the mat files.
% The command whos will show the files loaded
whos
% The open command will then open the file A, B and M
open A;
open B;
open M;
%The texture features for both the original and despeckled images and the quality evaluation
%metrics can now be manipulated or saved elsewhere.
```

References

[1] B. Fetics et al., "Enhancement of contrast echocardiography by image variability analysis," *IEEE IEEE Trans. Med. Imag.*, vol. 20, no. 11, pp. 1123–1130, Nov. 2001. DOI: 10.1109/42.963815. 2, 10, 11, 126

[2] C.I. Christodoulou, C. Loizou, C.S. Pattichis, M. Pantziaris, E. Kyriakou, M.S. Pattichis, C.N. Schizas, and A. Nicolaides,"Despeckle filtering in ultrasound imaging of the carotid artery," *Second Joint EMBS/BMES Conference*,Houston, TX, USA, pp. 1027–1028, Oct. 23–26, 2002. DOI: 10.2200/S00116ED1V01Y200805ASE001. 2, 18, 19, 124

[3] K. Abd-Elmoniem, A.-B. Youssef, and Y. Kadah, "Real-time speckle reduction and coherence enhancement in ultrasound imaging via nonlinear anisotropic diffusion," *IEEE Trans. Biomed. Eng.*, vol. 49, no. 9, pp. 997–1014, Sept. 2002. DOI: 10.1109/TBME.2002.1028423. 10, 11, 16, 17, 20, 41, 42, 86, 92, 94, 124, 126

[4] J.E. Wilhjelm, M.S. Jensen, S.K. Jespersen, B. Sahl, and E. Falk, "Visual and quantitative evaluation of selected image combination schemes in ultrasound spatial compound scanning," *IEEE Trans. Med. Imag.*, vol. 23, no. 2, pp. 181–190, 2004. DOI: 10.1109/TMI.2003.822824. 2, 10

[5] A. Achim, A. Bezerianos and P. Tsakalides, "Novel Bayesian multiscale method for speckle removal in medical ultrasound images," *IEEE Trans. Med. Imag.*, vol. 20, no. 8, pp. 772–783, 2001. DOI: 10.1109/42.938245. 2, 11, 16, 17, 41, 97, 124, 126

[6] C. Christodoulou, C. Pattichis, M. Pantziaris, and A. Nicolaides, "Texture-Based Classification of Atherosclerotic Carotid Plaques," *IEEE Trans. Med. Imag.*, vol. 22, no. 7, pp. 902–912, 2003. DOI: 10.1109/TMI.2003.815066. 2

[7] T. Elatrozy, A. Nicolaides, T. Tegos, A. Zarka, M. Griffin, and M. Sabetai, "The effect of B-mode ultrasonic image standardization of the echodensity of symptomatic and asymptomatic carotid bifurcation plaque," *Int. Angiology*, vol. 17: 179–186, no. 3, Sept. 1998. 2, 10, 12, 17, 20, 23, 34, 41, 57, 58, 79, 86, 124, 125, 126

[8] C.P. Loizou, C.S. Pattichis, M. Pantziaris, T. Tyllis, and A. Nicolaides, "Quantitative quality evaluation of ultrasound imaging in the carotid artery," *Med. Biol. Eng. & Comput.*," vol. 44, no. 5, pp. 414–426, 2006. DOI: 10.1007/s11517-006-0045-1. 10, 20, 34, 38

[9] C.P. Loizou, C.S. Pattichis, M. Pantziaris, and A. Nicolaides, "An integrated system for the segmentation of atherosclerotic carotid plaque," *IEEE Trans. Inform. Techn. Biomed.*," vol. 11, no. 5, pp. 661–667, Nov. 2007. DOI: 10.1109/TITB.2006.890019. 10, 17, 34, 129

[10] C.I. Christodoulou, S.C. Michaelides, and C.S. Pattichis, "Multi-feature texture analysis for the classification of clouds in satellite imagery," *IEEE Trans. Geoscience & Remote Sens.*, vol. 41, no. 11, pp. 2662–2668, Nov. 2003. DOI: 10.1109/TGRS.2003.815404.

[11] C.P. Loizou, C.S. Pattichis, C.I. Christodoulou, R.S.H. Istepanian, M. Pantziaris, and A. Nicolaides "Comparative evaluation of despeckle filtering in ultrasound imaging of the carotid artery," *IEEE Trans. Ultr. Fer. Freq. Contr.*, vol. 52, no. 10, pp. 1653–1669, 2005. DOI: 10.1109/TUFFC.2005.1561621.

[12] C.P. Loizou, V. Murray, M.S. Pattichis, M. Pantziaris, A.N. Nicolaides, and C.S. Pattichis, "Despeckle filtering for multiscale amplitude-modulation frequency-modulation (AM-FM) texture analysis of ultrasound images of the intima-media complex," *Int. J. Biomed. Imag.*, vol. 2014, Art. ID. 518414, 13 pages, 2014. DOI: 10.1155/2014/518414. 20, 26, 79

[13] C.P. Loizou, C. Theofanous, M. Pantziaris, and T. Kasparis, "Despeckle filtering software toolbox for ultrasound imaging of the common carotid artery," *Comput. Meth. & Progr. Biomed.*, vol. 114, no. 1, pp. 109–124, 2014. DOI: 10.1016/j.cmpb.2014.01.018. 2, 11, 12, 26, 31, 32, 33, 34, 41, 42, 79, 80, 81, 83, 134, 135

[14] C.P. Loizou, S. Petroudi, C.S. Pattichis, M. Pantziaris, and A.N. Nicolaides, "An integrated system for the segmentation of atherosclerotic carotid plaque in ultrasound video," *IEEE Trans. Ultras. Ferroel. Freq. Contr.*, vol. 61, no. 1, pp. 86–101, 2014. DOI: 10.1109/TUFFC.2014.6689778. 2, 10, 11, 23, 26

[15] C.P. Loizou, C. Theofanous, M. Pantziaris, T. Kasparis, P. Christodoulides, A.N. Nicolaides, and C.S. Pattichis, "Despeckle filtering toolbox for medical ultrasound video," *Int. J. Monitoring & Surveill. Technol. Resear.(IJMSTR): Special issue Biomed. Monitor. Technol.*, vol. 4, no. 1, pp. 61–79, 2013. DOI: 10.4018/ijmstr.2013100106. 2, 11, 12, 20, 26, 34, 37, 134, 135

[16] K.T. Dussik, "On the possibility of using ultrasound waves as a diagnostic aid," *Neurol. Psychiat.*, vol. 174, pp. 153–168, 1942. DOI: 10.1007/BF02877929. 2

[17] A. Kurjak, "Ultrasound scanning - Prof. Ian Donald (1910–1987)," *Eur. Journal Obstet. Gynecol. Reprod. Biol.*, vol. 90, no. 2, pp. 187-189, Jun. 2000. DOI: 10.1016/S0301-2115(00)00270-0. 2

[18] S.-M. Wu, Y.-W. Shau, F.-C. Chong, and F.-J. Hsieh, "Non-invasive assessment of arterial dimension waveforms using gradient-based Hough transform and power Doppler ultrasound imaging," *Journal of Med. Biol. Eng. & Comp.*, vol. 39, pp. 627–632, 2001. DOI: 10.1007/BF02345433. 4

[19] R. Gonzalez and R. Woods, *Digital image processing*, Second Edition, Prentice-Hall Inc., 2002. 4, 17, 19, 35, 68, 69

[20] W.R. Hedrick and D.L. Hykes, "Image and Signal Processing in Diagnostic Ultrasound Imaging," *J. Diagnostic Med. Sonogr.*, vol. 5, no. 5, pp. 231–239, 1989. DOI: 10.1177/875647938900500502. 7, 8

[21] F.J. Polak, *Doppler Sonography: An Overview, In Peripheral Vascular Sonography: A Practical Guide*, Baltimore USA: Williams and Wilkins, 1992. 7, 8, 10

[22] A Philips Medical System Company, "Comparison of image clarity, SonoCT real-time compound imaging versus conventional 2D ultrasound imaging," *ATL Ultrasound, Report*, 2001. 8, 10

[23] J.-C. Tradif and H. Lee, "Applications of ultravasular ultrasound in cardiology, What's new in cardiovascular ultrasound imaging," J. Reiber and E. van der Wall, Eds., pp. 133–148, Dordrecht: Kluwer Academic Publisher, 1998. 9

[24] M. Cilingiroglu, A. hakeem, M. Feldman, and M. Wholey, "Optical coherence tomography imaging in asymptomatic patients with carotid artery stenosis," *Cardiovascular Revascularization in Medicine*, vol. 14, no. 1, pp. 53–56, 2012. DOI: 10.1016/j.carrev.2012.09.004. 9, 85

[25] C.P. Loizou, C.S. Pattichis, M. Pantziaris, T. Tyllis, and A. Nicolaides, "Snakes based segmentation of the common carotid artery intima media," *Med. Biol. Eng. & Comput.*," vol. 45, no. 1, pp. 35–49, Jan. 2007. DOI: 10.1007/s11517-006-0140-3. 10, 17, 34, 115, 116, 117, 118, 129

[26] C.S. Pattichis, C. Christodoulou, E. Kyriakou, M. Pantziaris, A. Nicolaides, M.S. Pattichis, and C.P. Loizou, "Ultrasound imaging of carotid atherosclerosis," in *Wiley encyclopaedia of Biomedical Engineering*, Ed. By M. Akay, Wiley, Hoboken: John Wiley & Sons, Inc., USA, 2006. DOI: 10.1002/9780471740360.ebs1322. 10

[27] P. Rerona and J. Malik, "Scale-space and edge detection using anisotropic diffusion," *IEEE Trans. Pattern Anal. & Mach. Intellig.*, vol. 12, no. 7, pp. 629–639, July 1990. DOI: 10.1109/34.56205. 11, 17, 20, 85, 86, 125

[28] X. Hao, S. Gao, and X. Gao, "A novel multiscale nonlinear thresholding method for ultrasonic speckle suppressing," *IEEE Trans. Med. Imag.*, vol. 18, no. 9, pp. 787–794, 1999. DOI: 10.1109/42.802756. 11, 16, 17, 20, 23, 124, 126, 127

[29] Y. Yongjian and S.T. Acton, "Speckle reducing anisotropic diffusion," *IEEE Trans. Image Proces.*, vol. 11, no. 11, pp. 1260–1270, Nov. 2002. DOI: 10.1109/TIP.2002.804276. 11, 12, 17, 20, 23, 41, 85, 87, 88, 124, 125, 126

[30] D. Lamont, L. Parker, M. White, N. Unwin et al., "Risk of cardiovascular disease measured by carotid intima-media thickness at age 49–51: life course study," *BMJ*, vol. 320, pp. 273–278, 29 Jan. 2000. DOI: 10.1136/bmj.320.7230.273. 10

[31] C.P. Loizou, T. Kasparis, P. Papakyriakou, L. Christodoulou, M. Pantziaris, and C.S. Pattichis, "Video segmentation of the common carotid artery intima-media complex," *12th Int. Conf. Bioinf. & Bioeng. Proc. (BIBE)*, Larnaca, Cyprus, Nov. 11–13, pages 4, 2012. DOI: 10.1109/BIBE.2012.6399728. 10, 11

[32] C.B. Burckhardt, "Speckle in ultrasound B-mode scans," *IEEE Trans. Sonics Ultrasonics*, vol. SU-25, no. 1, pp. 1–6, 1978. DOI: 10.1109/T-SU.1978.30978. 11, 12, 17, 19, 20, 42, 86

[33] R.F. Wagner, S.W. Smith, J.M. Sandrik, and H. Lopez, "Statistics of speckle in ultrasound B-scans," *IEEE Trans. Sonics Ultrasonics*, vol. 30, pp. 156–163, 1983. DOI: 10.1109/T-SU.1983.31404. 11, 14, 16

[34] J.W. Goodman, "Some fundamental properties of speckle," *J. Optical Society of America*, vol. 66, no. 11, pp. 1145–1149, 1976. DOI: 10.1364/JOSA.66.001145. 11, 12, 14, 17

[35] R.W. Prager, A.H. Gee, G.M. Treece, and L. Berman, "Speckle detection in ultrasound images using first order statistics." University of Cambridge, Dept. of Engineering, GUED/F-INFENG/TR 415, pp. 1–17, July 2002. 11, 12, 42

[36] J.S. Lee, "Speckle analysis and smoothing of synthetic aperture radar images," *Computer Graph. Image Proces.*, vol. 17, pp. 24–32, 1981. DOI: 10.1016/S0146-664X(81)80005-6. 17, 18, 19, 41, 87, 124, 127

[37] C.P. Loizou, T. Kasparis, P. Christodoulides, C. Theofanous, M. Pantziaris, E. Kyriakou, and C.S. Pattichis, "Despeckle filtering in ultrasound video of the common carotid artery," *12th Int. Conf. Bioinf. & Bioeng. Proc. (BIBE)*, Larnaca, Cyprus, Nov. 11–13, pages 4, 2012. DOI: 10.1109/BIBE.2012.6399756. 11, 20, 26, 31, 42

[38] J.S. Lee, "Digital image enhancement and noise filtering by using local statistics," *IEEE Trans. Pattern Analysis & Machine Intellig.*, PAMI-2, no. 2, pp. 165–168, 1980. DOI: 10.1109/TPAMI.1980.4766994. 11, 17, 18, 19, 23, 41, 87, 124, 125

[39] M. Black, G. Sapiro, D. Marimont, and D. Heeger, "Robust anisotropic diffusion," *IEEE Trans. Image Proces.*, vol. 7, no. 3, pp. 421–432, March 1998. DOI: 10.1109/83.661192. 17, 85, 86, 124

[40] V. Dutt, "Statistical analysis of ultrasound echo envelope," Ph.D. dissertation, Mayo Graduate School, Rochester, MN, 1995. 11, 12, 14, 15, 16, 26, 37, 42, 124

[41] M. Insana et al., "Progress in quantitative ultrasonic imaging," *SPIE Vol. 1090 Medical Imaging III, Image Formation*, pp. 2–9, 1989. DOI: 10.1117/12.953184. 12, 17, 20

[42] J.C. Dainty, *Laser speckle and related phenomena*, Springer-Verlag, Berlin Heidelberg, New York, 1974. 12, 13, 14, 37

[43] J.M. Thijssen, B.J. Oosterveld, P.C. Hartman et al., "Correlations between acoustic and texture parameters from RF and B-mode liver echograms," *Ultrasound Med. Biol.*, vol. 19, pp. 13–20, 1993. DOI: 10.1016/0301-5629(93)90013-E. 12

[44] L. Busse, T.R. Crimmins, and J.R. Fienup, "A model based approach to improve the performance of the geometric filtering speckle reduction algorithm," *IEEE Ultrasonic Symposium*, pp. 1353–1356, 1995. DOI: 10.1109/ULTSYM.1995.495807. 14, 17, 18, 19, 23, 72

[45] J.U. Quistgaard, "Signal acquisition and processing in medical diagnostic ultrasound," *IEEE Signal Proces. Magazine*, vol. 14, no. 1, pp. 67–74, Jan. 1997. DOI: 10.1109/79.560325. 15

[46] H. Paul and H.P. Schwann, "Mechanism of absorption of ultrasound in liver tissue," *J. Acoustical Society America*, vol. 50, pp. 692, 1971. DOI: 10.1121/1.1912685. 14, 17, 18, 41

[47] V.S. Frost, J.A. Stiles, K.S. Shanmungan, and J.C. Holtzman, "A model for radar images and its application for adaptive digital filtering of multiplicative noise," *IEEE Trans. Pattern Analysis & Machine Intellig.*, vol. 4, no. 2, pp. 157–165, 1982. DOI: 10.1109/TPAMI.1982.4767223. 17, 18, 19, 20, 42, 69, 87, 124, 125

[48] J.S. Lee, "Refined filtering of image noise using local statistics," *Computer Graph. & Image Proces.*, vol. 15, pp. 380–389, 1981. DOI: 10.1016/S0146-664X(81)80018-4. 17, 18, 19, 41, 87, 124

[49] D.T. Kuan, A.A. Sawchuk, T.C. Strand, and P. Chavel, "Adaptive restoration of images with speckle," *IEEE Trans. Acoustic Speech & Signal Processing*, vol. ASSP-35, pp. 373–383, 1987. DOI: 10.1109/TASSP.1987.1165131. 17

[50] D.T. Kuan and A.A. Sawchuk, "Adaptive noise smoothing filter for images with signal dependent noise," *IEEE Trans. Pattern Analysis & Mach. Intellig.*, vol. PAMI-7, no. 2, pp. 165–177, 1985. DOI: 10.1109/TPAMI.1985.4767641. 17, 18, 19, 20, 41, 124

[51] C.P. Loizou , C. Christodoulou, C.S. Pattichis, R. Istepanian, M. Pantziaris, and A. Nicolaides, "Speckle reduction in ultrasound images of atherosclerotic carotid plaque," *DSP*

2002, Proc. IEEE 14th Int. Conf. Digital Signal Proces., Santorini-Greece, pp. 525–528, July 1–3, 2002. DOI: 10.1109/ICDSP.2002.1028143. 17, 18, 19, 20, 78, 124, 126

[52] X. Zong, A. Laine, and E. Geiser, "Speckle reduction and contrast enhancement of echocardiograms via multiscale nonlinear processing," *IEEE Trans. Med. Imag.*, vol. 17, no. 4, pp. 532–540, 1998. DOI: 10.1109/42.730398. 17, 97, 126, 127

[53] M. Karaman, M. Alper Kutay, and G. Bozdagi, "An adaptive speckle suppression filter for medical ultrasonic imaging," *IEEE Trans. Med. Imag.*, vol. 14, no. 2, pp. 283–292, 1995. DOI: 10.1109/42.387710. 17, 126, 127

[54] S.M. Ali and R.E. Burge, "New automatic techniques for smoothing and segmenting SAR images," *Signal Processing*, North-Holland, vol. 14, pp. 335–346, 1988. DOI: 10.1016/0165-1684(88)90092-8. 17, 70, 124

[55] A. Baraldi and F. Pannigianni, "A refined gamma MAP SAR speckle filter with improved geometrical adaptivity," *IEEE Trans. Geoscience & Remote Sensing*, vol. 33, no. 5, pp. 1245–1257, Sep. 1995. DOI: 10.1109/36.469489.

[56] E. Trouve, Y. Chambenoit, N. Classeau, and P. Bolon, "Statistical and operational performance assessment of multi-temporal SAR image filtering," *IEEE Trans. Geosc. & Remote Sens.*, vol. 41, no. 11, pp. 2519–2539, 2003. DOI: 10.1109/TGRS.2003.817270. 17, 124

[57] H.-L. Eng and K.-K. Ma, "Noise adaptive soft-switching median filter," *IEEE Trans. Image Process.*, vol. 10, no. 2, pp. 242–251, 2001. DOI: 10.1109/83.902289. 17, 126, 127

[58] S. Solbo and T. Eltoft, "Homomorphic wavelet based-statistical despeckling of SAR images," *IEEE Trans. Geosc. Remote Sensing*, vol. 42, no. 4, pp. 711–721, 2004. DOI: 10.1109/TGRS.2003.821885. 17

[59] J. Saniie, T. Wang, and N. Bilgutay, "Analysis of homomorphic processing for ultrasonic grain signal characterization," *IEEE Trans. Ultr., Fer. & Freq. Contr.*, vol. 3, pp. 365–375, 1989. DOI: 10.1109/58.19177. 17, 18, 19, 78

[60] T. Huang, G. Yang, and G. Tang, "A fast two-dimensional median filtering algorithm," *IEEE Trans. Acoustics, Speech & Sign. Proces.*, vol. 27, no. 1, pp. 13–18, 1979. DOI: 10.1109/TASSP.1979.1163188. 17, 19, 20, 67, 85, 124

[61] J. Weickert, B. Romery, and M. Viergever, "Efficient and reliable schemes for non-linear diffusion filtering," *IEEE Trans. Image Proces.*, vol. 7, pp. 398–410, 1998. DOI: 10.1109/83.661190. 17, 20, 56, 85

[62] S. Jin, Y. Wang, and J. Hiller, "An adaptive non-linear diffusion algorithm for filtering medical images," *IEEE Trans. Inf. Techn. Biomed.*, vol. 4, no. 4, pp. 298–305, Dec. 2000. DOI: 10.1109/4233.897062. 17

[63] M. Larsson, B. Heyde, F. Kremer, L.-A. Brodin, and J. D'hooge, "Ultrasound speckle tracking for radial, longitudinal and circumferential strain estimation of the carotid artery – An in vitro validation via sonomicrometry using clinical and high-frequency ultrasound," *Ultrasonics*", accepted. DOI: 10.1016/j.ultras.2014.09.005. 17, 18

[64] J. D'hooge, E. Konofagou, F. Jamal, A. Heimdal, L. Barrios, B. Bijnens, J. Thoen, F. Van de Werf, G. Sutherland, and P. Suetens, "Two-dimensional ultrasonic strain rate measurement of the human heart in Vivo," *IEEE Trans. Ultras. Fer. Freq. Contr.*, vol. 49, pp. 281–286, 2002. DOI: 10.1109/58.985712. 17

[65] C.L. de Korte, H.H. Hansen, A.F. and van der Steen, "Vascular ultrasound for atherosclerosis imaging," *Interf. Focus*, vol. 1, pp. 565–575, 2001. DOI: 10.1098/rsfs.2011.0024. 17

[66] M. Catalano, A. Lamberti-Castronuovo, A. Catalano, D. Filocamo, and C. Zimbalatti, "Two-dimensional speckle-tracking strain imaging in the assessment of mechanical properties of carotid arteries: feasibility and comparison with conventional markers of subclinical atherosclerosis," *Eur. J. Echocardiogr.*, vol. 12, pp. 528–535, 2001. DOI: 10.1093/ejechocard/jer078. 17, 18

[67] M. Cinthio, Å. Rydén Ahlgren, T. Jansson, A. Eriksson, H.W. Persson, and L. Kjell, "Evaluation of an ultrasonic echo-tracking method for measurements of arterial wall movements in two dimensions," *IEEE Trans. Ultr. Fer. Freq. Contr.*, vol. 52, pp. 1300–1311, 2005. DOI: 10.1109/TUFFC.2005.1509788. 18

[68] M. Larsson, F. Kremer, P. Claus, T. Kuznetsova, L.-Å. Brodin, and J. D'hooge, "Ultrasound-based Radial and Longitudinal Strain Estimation of the Carotid Artery: a feasibility study," *IEEE Trans. Ultr. Fer. Freq. Contr.*, vol. 58, pp. 2244–2251, 2011. DOI: 10.1109/TUFFC.2011.2074. 18

[69] R. Bernardes, C. Mduro, P. Serranho, A. Araujo, S. Barbeiro, and J. Cunha-Vaz, "Improved adaptive complex diffusion despeckling filter," *Opt. Express*, vol. 18, pp. 24048–24059, 2010. DOI: 10.1364/OE.18.024048. 18, 20, 69, 94

[70] S. Zhong and V. Cherkassky, "Image denoising using wavelet thresholding and model selection," *Proc. of IEEE Int. Conf. Image Proces.*, Vancouver, Canada, pp. 1–4, Nov. 2000. DOI: 10.1109/ICIP.2000.899365. 20, 23, 42, 97

[71] D.L. Donoho, "Denoising by Soft Thresholding," *IEEE Trans. Inform. Theory*, vol. 41, pp. 613–627, 1995. DOI: 10.1109/18.382009. 20, 23, 26, 97, 126, 127

[72] T.W. Chan, O.C. Au, T.S. Chong, and W.S. Chau, "A novel content-adaptive video denoising filter," *Proc. IEEE Vision, Imag. & Sign. Proces.*, vol. 2, no. 2, pp. 649–652, 2005. DOI: 10.1109/ICASSP.2005.1415488. 20

[73] K. Dabov, A. Foi, and K. Egiazarian, "Video denoising by sparse 3D transform-domain collaborative filtering," *Proc. 15th Eur. Sign. Proc. Conf.*, pp. 1–5, 2007.

[74] M. Maggioni, G. Boracchi, A. Foi, and K. Engiazarian, "Video denoising, deblocking and enhancement through separable 4-D nonlocal spatiotemporal transforms," *IEEE Trans. Imag. Proc.*, vol. 21, no. 9, pp. 3952–3966, 2012. DOI: 10.1109/TIP.2012.2199324.

[75] D. Rusanovskyy, K. Dabov, and K. Egiazarian, "Moving-window varying size 3D transform-based video denoising," *Proc. Int. Workshop on Video Proc. & Quality Metrics*, 1–4, 2006.

[76] V. Zlokolica, W. Philips, and van de Ville, "Robust non-linear filtering for video processing," *IEEE Proc. Vision, Imag. & Sign. Proces.*, vol. 2, no. 2, pp. 571–574, 2002. DOI: 10.1109/ICDSP.2002.1028154.

[77] V. Zlokolica, A. Pizurica, and W. Philips, "Recursive temporal denoising and motion estimation of video," *Int. Conf. on Image Proces.*, vol. 3, no. 3, pp. 1465–1468, 2008. DOI: 10.1109/ICIP.2004.1421340. 20

[78] C.P. Loizou and C.S. Pattichis, *Despeckle filtering algorithms and Software for Ultrasound Imaging. Synthesis lectures on algorithms and software for engineering*, Ed. Morgan & Claypool Publishers, San Rafael, CA, USA, 2008. DOI: 10.2200/S00116ED1V01Y200805ASE001. 20, 41, 42, 79

[79] M. Nagao and T. Matsuyama, "Edge preserving smoothing," *Computer Graph. & Image Proces.*, vol. 9, pp. 394–407, 1979. DOI: 10.1016/0146-664X(79)90102-3. 19, 59

[80] T. Greiner, C.P. Loizou, M. Pandit, J. Mauruschat, and F.W. Albert, "Speckle reduction in ultrasonic imaging for medical applications," *Proc. of the ICASSP91, 1991 Int. Conf. Acoustic Signal Speech Processing*, Toronto Canada, May 14–17, pp. 2993-2996, 1991. DOI: 10.1109/ICASSP.1991.151032. 19

[81] A. Nieminen, P. Heinonen, Y. Neuvo, "A new class of detail-preserving filters for image processing," *IEEE Trans. Pattern Anal. Mach. Intell.*, vol. 9, pp. 74–90, 1987. DOI: 10.1109/TPAMI.1987.4767873. 19, 79

[82] M. Kuwahara, K. Hachimura, S. Eiho, and M. Kinoshita, *Digital processing of biomedical images*, Plenum. Pub. Corp., Ed. K. Preston and M. Onoe, pp. 187–203, 1976. 19, 80

[83] A. Buades, B. Coll, and J.-M. Morel, "Nonlocal image and movie denosing," *Int. J. Comput. Vis.*, vol. 76, pp. 123–139, 2008. DOI: 10.1007/s11263-007-0052-1. 19, 82, 83

[84] A.M. Wink and J.B.T.M. Roerdink, "Denoising functional MR images: A comparison of wavelet denoising and Gaussian smoothing," *IEEE Trans. Med. Imag.*, vol. 23, no. 3, pp. 374–387, 2004. DOI: 10.1109/TMI.2004.824234. 20

[85] N. Rougon and F. Preteux, "Controlled anisotropic diffusion," *Conf. on Nonlinear Image Processing VI*, IS&T/SPIE Symposium on Electronic Imaging, Science and Technology, San Jose, California, pp. 1–12, 5–10 Feb. 1995. DOI: 10.1117/12.205235. 20, 126

[86] F.N.S. Medeiros, N.D.A. Mascarenhas, R.C.P Marques, and C.M. Laprano, "Edge preserving wavelet speckle filtering," *5th IEEE Southwest Symp. Image Anal. & Interpr.*, Santa Fe, New Mexico, pp. 281–285, 7–9 April 2002. DOI: 10.1109/IAI.2002.999933. 20

[87] P. Moulin, "Multiscale image decomposition and wavelets," in *Handbook of Image & Video Processing*, Ed. by A. Bovik, Academic Press, pp. 289–300, 2000. 23

[88] P. Scheunders, "Wavelet thresholding of multivalued images," *IEEE Trans. Image Proces.*, vol. 13, no. 4, pp. 475–483, 2004. DOI: 10.1109/TIP.2004.823829. 23

[89] Christos P. Loizou, PhD thesis, "Ulytrasound image processing for the evaluation of the risk of stroke," Kingston University, UK, 2005. 29, 97

[90] J.T. Bushberg, J. Anthony Seibert, E.M. Leidholdt, Jr., and J.M. Boone, *The essential physics of medical imaging*, Lippincott Williams & Wilkins, 2002. 29

[91] J. Stoitsis, S. Golemati, V. Koropouli, and K.S. Nikita, "Simulating dynamic B-mode ultrasound image data of the common carotid artery," *IEEE Int. Work. Imag. Synth. & Techn.*, pages 4, 2008. DOI: 10.1109/IST.2008.4659958. 31

[92] E. Krupinski, H. Kundel, P. Judy, and C. Nodine, "The medical image perception society, key issues for image perception research," *Radiology*, vol. 209, pp. 611–612, 1998. DOI: 10.1148/radiology.209.3.9844649. 32, 39

[93] Z. Wang, A. Bovik, H. Sheikh, and E. Simoncelli, "Image quality assessment: From error measurement to structural similarity," *IEEE Trans. Image Proces.*, vol. 13, no. 4, pp. 600–612, Apr. 2004. DOI: 10.1109/TIP.2003.819861. 34, 35, 36, 39

[94] A. Ahumada and C. Null, "Image quality: A multidimensional problem," in *Digital images and human vision*, Ed. A.B. Watson, Bradford Press: Cambridge Mass, pp. 141–148, 1993. 34

[95] E.A. Fedorovskaya, H. De Ridder, and F.J. Blomaert, "Chroma variations and perceived quality of colour images and natural scenes," *Color Research & Application*, vol. 22, no. 2, pp. 96–110, 1997. DOI: 10.1002/(SICI)1520-6378(199704)22:2%3C96::AID-COL5%3E3.0.CO;2-Z. 34

[96] G. Deffner, "Evaluation of display image quality: Experts vs. non-experts," *Symposium Society for Information and Display Digest*, vol. 25, pp. 475–478, 1994. 34

[97] T.J. Chen, K.S. Chuang, Jay Wu, S.C. Chen, I.M. Hwang, and M.L. Jan, "A novel image quality index using Moran I statistics," *Physics Med. & Biol.*, vol. 48, pp. 131–137, 2003. DOI: 10.1088/0031-9155/48/8/402. 35

[98] S. Winkler, "Vision models and quality metrics for image processing applications," PhD, University of Lausanne-Switzerland, Dec. 21, 2000. 35, 39

[99] D. Sakrison, "On the role of observer and a distortion measure in image transmission," *IEEE Trans. Communic.*, vol. 25, pp. 1251–1267, Nov. 1977. DOI: 10.1109/TCOM.1977.1093773. 36

[100] Z. Wang and A. Bovik, "A Universal quality index," *IEEE Sign. Proces. Letters*, vol. 9, no. 3, pp. 81–84, March 2002. DOI: 10.1109/97.995823. 36

[101] V.S. Vora, A.C. Suthar, Y.N. Makwana, and S.J. Davda, "Analysis of compresed image quality assessments," *Int. Journal of Advanced Engineering and Applications*, vol. 1, pp. 225–229, 2010. 37, 38

[102] A. Pommert and K. Hoehne, "Evaluation of image quality in medical volume visualization: The state of the art," Takeyoshi Dohi, Ron Kikinis (Eds.): in *Medical image computing and computer-assisted intervention, Proc. MICCAI, 2002, Part II, Lecture Notes in Computer Science 2489*, pp. 598–605, Springer Verlag, Berlin 2002. DOI: 10.1007/3-540-45787-9_75. 39

[103] M. Eckert, "Perceptual quality metrics applied to still image compression," Canon information systems research, Faculty of engineering, Univ. Of Technology, Sydney, Australia, pp. 1–26, 2002. DOI: 10.1016/S0165-1684(98)00124-8. 39

[104] A. Efros, T. Leung, "Texture synthesis by non parametric sampling," *Proc. Int. Conf. Computer Vision*, vol. 2, pp. 1033–1038, 1999. DOI: 10.1109/ICCV.1999.790383. 82

[105] Y. Zhan, M. Ding, L. Wu, and X. Zhang, "Nonlocal means method using wight refining for despeckling for ultrasound images," *Sign. Proc.*, vol. 103, pp. 201–213, 2014. DOI: 10.1016/j.sigpro.2013.12.019. 82, 83

[106] A. Buades, B. Coll, and J. Morel, "On image denoising methods," Technical Report, CMLA, 2004–15. 83

[107] G. Gilboa, N. Sochen, and Y.Y. Zeevi, "Image enhancement and denoising by complex diffusion processes," *IEEE Trans. Pattern Anal. Mach. Intell.*, vol. 26, no. 8, pp. 1020–1036, 2004. DOI: 10.1109/TPAMI.2004.47. 94

[108] R.N. Czerwinski, D.L. Jones, and W.D. O'Brien, "Detection and boundaries in speckle images-Application to medical ultrasound," *IEEE Trans. Med. Imag.*, vol. 18, no. 2, pp. 126–136, Feb. 1999. DOI: 10.1109/42.759114. 126

[109] A.M. Wink and J.B.T.M. Roerdink, "Denoising functional MR images: A comparison of wavelet denoising and Gaussian smoothing," *IEEE Trans. Med. Imag.*, vol. 23, no. 3, pp. 374–387, 2004. DOI: 10.1109/TMI.2004.824234. 126

[110] S. Jin, Y. Wang, and J. Hiller, "An adaptive non-linear diffusion algorithm for filtering medical images," *IEEE Trans. Inform. Technol. Biomed.*, vol. 4, no. 4, pp. 298–305, Dec. 2000. DOI: 10.1109/4233.897062. 126

[111] V. Damerjian, O. Tankyevych, N. Souag, and E. Petit, "Speckle characterisation methods in ultrasound images-A review," *Innovat. & Resear. Biomed. Eng. (IRBM)*, vol. 35, pp. 202–213, 2014. DOI: 10.1016/j.irbm.2014.05.003. 5, 127

[112] J.A. Noble, "Ultrasound image segmentation and tissue characterization," *Proc. Inst. Mech. Eng.*, vol. 224, no. 2, pp. 307–316, 2010. DOI: 10.1243/09544119JEIM604. 127

[113] C.P. Loizou, "A review on ultrasound common carotid artery image and video segmentation techniques," *Med. Biol. Eng. Comput.*, vol. 52, no. 12, pp. 1073–1093, 2014. DOI: 10.1007/s11517-014-1203-5. 129

[114] R.J. Housden, A.H. Gee, G.M. Treece, and R.W. Prager, "Sensorless reconstruction of unconstrained freehand 3D ultrasound data," *Ultrasound Med. Biol.*, vol. 33, no. 3, pp. 408–419, 2007. DOI: 10.1016/j.ultrasmedbio.2006.09.015. 127

[115] J.C. Seabra, F. Ciompi, O. Pujol, J. Mauri, P. Radeva, and J. Sanches, "Rayleighmixture model for plaque characterization in intravascular ultrasound," *IEEE Biomed. Eng.*, vol. 58, no. 5, pp. 1314–1324, 2001. DOI: 10.1109/TBME.2011.2106498. 127

[116] M.A.H. Khan, "3D reconstruction of ultrasound images," MSc in vision and robotics, University of Burgundy, University of Girona and University of Heriot Watt., 2008. 127

[117] J. Revell, M. Mirmehdi, and D. McNally, "Computer vision elastography:speckle adaptive motion estimation for elastography using ultrasoundsequences," *IEEE Med. Imaging*, vol. 24, no. 6, pp. 755–66, 2005. DOI: 10.1109/TMI.2005.848331. 127

[118] H. Geyer, G. Caracciolo, H. Abe, S. Wilansky, S. Carerj, F. Gentile et al., "Assessment of myocardial mechanics using speckle tracking echocardi-ography: fundamentals and clinical applications," *J. Am. Soc. Echocardiogr.*, vol. 23, no. 4, pp. 351–69, 2010. DOI: 10.1016/j.echo.2010.02.015. 127

[119] M. Alessandrini, PhD thesis, "Statistical methods for analysis and processing of medical ultrasound: applications to segmentation and restoration." Universita di Bologna, 2011. 127

[120] F. Destrempes, G. Soulez, M.-F. Giroux, J. Meunier, and G. Cloutier, "Segmentation of plaques in sequences of ultrasonic B-mode images of carotidarteries based on motion estimation and Nakagami distributions," *IEEE Int. Ultrasonics Symposium (IUS)*, pp. 2480–24833, 2009. DOI: 10.1109/ULTSYM.2009.5441741. 127

[121] N. Paragios and R. Deriche, "Geodesic active contours and level sets for the detection and tracking of moving objects," *IEEE Trans. Pattern. Anal. Mach. Intell.*, vol. 22, no. 3, pp. 266–80, 2000. DOI: 10.1109/34.841758. 127

[122] Y.Y. Boykov and M.-P. Jolly, "Interactive graph cuts for optimal boundary & region segmentation of objects in ND images," *Computer Vision, ICCV 2001. Proc. 8th IEEE Int. Conf.*, pp. 105–112, 2001. DOI: 10.1109/ICCV.2001.937505. 127

[123] M. Oezkan, A. Erdem, M. Sezan, and A. Tekalp, "Effcient multi-frame Wiener restoration of blurred and noisy image sequences," *IEEE Trans. Image Proces.*, vol. 1, pp. 453–476, Oct. 1992. DOI: 10.1109/83.199916. 128

[124] P.M.B. Van Roosmalen, S.J.P. Westen, R.L. Lagendijk, and J. Biemond, "Noise reduction for image sequences using an oriented pyramid threshold technique," *IEEE Int. Conf. Image Proces.*, vol. 1, pp. 375–378, 1996. DOI: 10.1109/ICIP.1996.559511. 128

[125] M. Vetterli, J. Kovacevic, *Wavelets and subband coding*, Prentice Hall, 1995. 128

[126] S. Winkler, *Digital video quality, Vision models and metrics*, John Wiley & Sons, 2005. DOI: 10.1002/9780470024065. 128, 131

[127] J.-H. Jung, K. Hong, and S. Yang, "Noise reduction using variance characteristics in noisy image sequence," *Int. Conf. Consumer Electronics*, pp. 213–214, 8–12 Jan. 2005. DOI: 10.1109/ICCE.2005.1429793. 128

[128] M. Bertalmio, V. Caselles, and A. Pardo, "Movie Denoising by average of warped lines," *IEEE Trans. Image Proces.*, vol. 16, no. 9, pp. 233–2347, 2007. DOI: 10.1109/TIP.2007.901821. 128

[129] B. Alp, P. Haavisto, T. Jarske, K. Oestaemoe, and Y. Neuro, "Median based algorithms for image sequence processing," *SPIE Visual Commun. & Image Proces.*, pp. 122–133, 1990. DOI: 10.1117/12.24175. 128, 131

[130] A. Panayides, M.S. Pattichis, C.S. Pattichis, C.P. Loizou, M. Pantziaris, and A. Pitsillides, "Atherosclerotic plaque ultrasound video encoding, wireless transmission, and quality assessment using H.264," *IEEE Trans. Inform. Tech. Biomed.*, vol. 15, no. 3, pp. 387–397, 2011. DOI: 10.1109/TITB.2011.2105882. 129, 131

[131] T. Painter and A.S. Spanias, "Perceptual Coding of Digital Audio," *Proc. IEEE*, vol. 88, no. 4, pp. 451–513, 2000. DOI: 10.1109/5.842996. 129, 131

[132] A.S. Spanias, *Digital Signal Processing; An Interactive Approach*, 2nd ed., 403 pages, Textbook with JAVA exercises, ISBN 978–1-4675-9892-, Lulu Press On-demand Publishers Morrisville, NC, May 2014.

[133] A.S. Spanias, "Speech Coding: A Tutorial Review," *Proc. IEEE*, pp. 1441–1582, vol. 82, no. 10, 1994. DOI: 10.1109/5.326413.

[134] J. Thiagarajan, K. Ramamurthy, P. Turaga, and A. Spanias, "Image Understanding using sparse representations, in *Synthesis Lectures on Image, Video, and Multimedia Processing*, 978–1627053594, 118 pages, Ed. Al Bovik, 2014. DOI: 10.2200/S00563ED1V01Y201401IVM015. 131

[135] C.P. Loizou and C.S. Pattichis, *Despeckle filtering for ultrasound imaging and video*, Volume II: Selected applicatios, Ed. Morgan & Claypool Publishers, CA, USA, 2015. DOI: 10.2200/S00116ED1V01Y200805ASE001. 132

Authors' Biographies

CHRISTOS P. LOIZOU

Christos P. Loizou was born in Cyprus on October 23, 1962, received his B.Sc. degree in electrical engineering, a Dipl-Ing (M.Sc.) degree in computer science and telecommunications from the University of Kaisserslautern, Kaisserslautern, Germany, and his Ph.D. degree on ultrasound image analysis of the carotid artery from the Department of Computer Science, Kingston University, London, UK, in 1986, 1990, and 2005, respectively. From 1996–2000, he was a lecturer in the Department of Computer Science, Higher Technical Institute, Nicosia, Cyprus. Since 2000, he has been at the Department of Computer Science, Intercollege, Cyprus and is now a campus program coordinator. Since 2005, he has also been an Adjunct Professor of Medical Image and video processing, in the Department of Electrical Engineering, and Computer Engineering and Informatics, Cyprus University of Technology, Cyprus. He has also been an Associated Researcher at the Institute of Neurology and Genetics, Nicosia, Cyprus since 2000. Dr. Loizou was a supervisor of a number of Ph.D. and B.Sc. students in the area of computer image analysis and telemedicine. He was involved in the research activity of several scientific Cypriot and European research projects and has authored or co-authored 28 referred journals, 54 conference papers, 3 books, and 10 book chapters in the fields of image and video analysis. His current research interests include medical imaging, signal, image and video processing, motion and video analysis, pattern recognition, and biosignal analysis in ultrasound, magnetic resonance imaging, and computer applications in medicine. He is a Senior Member of the IEEE, serves as a reviewer in many IEEE Transactions and other journals and is a chair and co-chair at many IEEE conferences. He lives in Limassol, Cyprus, with his wife and children, a boy and a girl.

CONSTANTINOS S. PATTICHIS

Constantinos S. Pattichis was born in Cyprus on January 30, 1959 and received his diploma as technician engineer from the Higher Technical Institute in Cyprus in 1979, a B.Sc. in electrical engineering from the University of New Brunswick, Canada, in 1983, the M.Sc. in biomedical engineering from the University of Texas at Austin, USA, in 1984, an M.Sc. in neurology from the University of Newcastle Upon Tyne, UK, in 1991, and his Ph.D. in electronic engineering from the University of London, UK, in 1992. He is currently a Professor with the Department of Computer Science of the University of Cyprus. His research interests include ehealth and mhealth, medical imaging, biosignal analysis, life sciences informatics, and intelligent systems. He has been involved in numerous projects in these areas funded by EU, the National Research Foundation of Cyprus, the INTERREG and other bodies, like the FI-STAR, GRANATUM, LINKED2SAFETY, MEDUCATOR, LONG LASTING MEMO-RIES, INTRAMEDNET, INTERMED, FUTURE HEALTH, AMBULANCE, EMER-GENCY, ACSRS, TELEGYN, HEALTHNET, IASIS, IPPOKRATIS, and others with a to-tal funding managed of more than 6 million Euros. He has published 90 refereed journal and 200 conference papers, and 27 chapters in books in these areas. He is Co-Editor of the books *M-Health: Emerging Mobile Health Systems*, and *Ultrasound and Carotid Bifurcation Atherosclero-sis* published by Springer in 2006 and 2012, respectively. He is co-author of the book *Despeckle Filtering Algorithms and Software for Ultrasound Imaging*, published by Morgan & Claypool Pub-lishers in 2008 and the revised second edition to be published in 2015. He was Guest Co-Editor of the Special Issues on *Atherosclerotic Cardiovascular Health Informatics*, *Emerging Health Telematics Applications in Europe*, *Emerging Technologies in Biomedicine*, *Computational Intelligence in Med-ical Systems*, and *Citizen Centered e-Health Systems in a Global Health-care Environment* of the IEEE Transactions on Information Technology in Biomedicine. He is General Chairman of the forthcoming *Medical and Biological Engineering and Computing Conference (MEDICON'2016)* and the *IEEE Region 8 Mediterranean Conference on Information Technology and Electrotechnology (MELECON'2016)*. He was General Co-Chairman of the *IEEE 12th International Conference on BioInformatics and BioEngineering (BIBE2012)*, *IEEE Information Technology in Biomedicine (ITAB09)*, *MEDICON'98, MELECON'2000*, and Program Co-Chair of *ITAB06* and the *4th In-ternational Symposium on Communications, Control and Signal Processing (ISCCSP 20010)*. More-over, he serves as an Associate Editor of the *IEEE Journal of Biomedical and Health Informatics*, on the Editorial Board of the *Journal of Biomedical Signal Processing and Control*, and as mem-ber of the IEEE EMBS Technical Committee on Information Technology for Health (since 2011). He served as Distinguished Lecturer of the IEEE EMBS (2013–2014), and an Associate Editor of the *IEEE Transactions on Information Technology in Biomedicine* (2000–2012) and the *IEEE Transactions on Neural* (2005–2007). He served as Chairperson of the Cyprus Association of Medical Physics and Biomedical Engineering (1996–1998), and the IEEE Cyprus Section (1998-2000). He is a Fellow of IET and a Senior Member of IEEE.

Printed in the United States
by Baker & Taylor Publisher Services